ACS SYMPOSIUM SERIES **316**

Formaldehyde Release from Wood Products

B. Meyer, EDITOR
University of Washington

B. A. Kottes Andrews, EDITOR
U.S. Department of Agriculture

Robert M. Reinhardt, EDITOR
U.S. Department of Agriculture

Developed from a symposium sponsored by
the Division of Cellulose, Paper, and Textile Chemistry
at the 189th Meeting
of the American Chemical Society,
Miami Beach, Florida,
April 28–May 3, 1985

American Chemical Society, Washington, DC 1986

Library of Congress Cataloging-in-Publication Data

Formaldehyde release from wood products.
  (ACS symposium series, ISSN 0097-6156; 316)

  "Based on a symposium sponsored by the Division of
Cellulose, Paper, and Textile Chemistry at the 189th
Meeting of the American Chemical Society."
  Includes bibliographies and index.

  1. Wood products—Formaldehyde release—
Congresses. 2. Urea-formaldehyde resins—Congresses.

  I. Meyer, Beat. II. Andrews, B. A. Kottes (Bethlehem
A. Kottes), 1936-   . III. Reinhardt, R. M.
IV. American Chemical Society. Cellulose, Paper, and
Textile Division. V. American Chemical Society.
Meeting (189th: Miami Beach, Fla.) VI. Series.

TS858.F67   1986   674'.8   86-14194
ISBN 0-8412-0982-0

ACS Symposium Series

M. Joan Comstock, *Series Editor*

FOREWORD

The ACS SYMPOSIUM SERIES was founded in 1974 to provide a medium for publishing symposia quickly in book form. The format of the Series parallels that of the continuing ADVANCES IN CHEMISTRY SERIES except that, in order to save time, the papers are not typeset but are reproduced as they are submitted by the authors in camera-ready form. Papers are reviewed under the supervision of the Editors with the assistance of the Series Advisory Board and are selected to maintain the integrity of the symposia; however, verbatim reproductions of previously published papers are not accepted. Both reviews and reports of research are acceptable, because symposia may embrace both types of presentation.

CONTENTS

PREFACE

THIS BOOK SUMMARIZES OUR CURRENT UNDERSTANDING of many problems related to measuring, abating, and understanding formaldehyde emission from wood products bonded with formaldehyde-based adhesive resins. It contains expanded and updated versions of selected papers presented at an ACS symposium, "Formaldehyde Release from Cellulose in Wood Products and Textiles." In addition, three chapters from participants who could not attend the meeting were added.

The first three chapters deal with particleboard, medium density fiberboard, hardwood plywood, and softwood plywood, the four most widely used wood panel products. Chapter four compares these products with other consumer products. Chapters five through seven explain the basic chemistry of formaldehyde with cellulose and wood components and provide a current understanding of the nature of liquid urea–formaldehyde adhesive resins. The next two chapters present new analytical methods that might become useful in the future. Chapters eight and eleven through sixteen explain the complex nature of the latent formaldehyde present in the products and its correlation to formaldehyde emission from wood products. Chapters fifteen and sixteen describe currently popular formaldehyde reduction methods. The last two chapters discuss the problems involved in reducing formaldehyde emission by regulating air levels or source emissions.

The editors thank all contributors for their excellent cooperation.

B. A. KOTTES ANDREWS
U.S. Department of Agriculture
New Orleans, LA 70179

BEAT MEYER
University of Washington
Seattle, WA 98195

ROBERT M. REINHARDT
U.S. Department of Agriculture
New Orleans, LA 70179

May 1, 1986

Formaldehyde Release from Wood Products: An Overview

B. Meyer and K. Hermanns

Chemistry Department, University of Washington, Seattle, WA 98195

Formaldehyde release from UF-bonded wood products has decreased by a factor of more than ten over the past 15 years. Today 90% of the entire U.S. production is capable of meeting the 0.4 ppm standard for manufactured housing at the time of sale. Since 1979 European products have been classified into three categories. Low emitting materials meeting 0.l ppm air levels currently account for about 20% of the European production. These low emitting products can be made by different methods: Using low F:U molar ratio resins, addition of urea to resin or wood furnish before resination, and post-treatment of hot board with ammonia or ammonia salts. Due to improved quality control, current products are now capable of meeting predictable emission performance criteria and, in most applications, they can be used in a traditional load ratio without air levels exceeding 0.l ppm under normal use conditions.

During the past forty years wood panel products bonded with formaldehyde derived resins have become increasingly popular and have replaced whole wood in almost every use. Thus, these products are now present as construction material and in furniture and cabinet work in almost every building. At the current load factors of 0.2 to 1 m^2 of product surface per 1 m^3 of indoor air volume even traces of residual, unreacted adhesive vapors are sufficient to cause noticeable indoor air concentrations and odors. Because of its high vapor pressure, formaldehyde is the most significant of these emitters.

Wood Products

The most widely used wood panel products are particleboard, softwood plywood, hardwood plywood, medium density fiberboard (MDF) and waferboard. The most common adhesive is urea-formaldehyde resin (UFR). Phenol-formaldehyde resins (PFR) are second in volume and melamine-formaldehyde resins (MFR) are a distant third. Recently,

0097–6156/86/0316–0001$06.00/0

some specialty products have been bonded with isocyanates. The
tendency to release residual formaldehyde differs significantly among
these products.

Particleboard and other products made with isocyanates emit only
little formaldehyde (1), but these adhesives are expensive and
require expensive manufacturing procedures. In contrast, phenolic
soft wood plywood is a well established product that is predominantly
used for exterior applications. It contains formaldehyde in
chemically strongly bonded form and also emits little formaldehyde,
as shown in a later chapter in this book. In fact, under almost all
common use conditions this type of board contributes not much more
formaldehyde than is already present in ambient air in many urban
areas. The same is true for waferboard, which has recently become
popular for replacing plywood. Likewise, phenolic particleboard
emits little formaldehyde, unless the phenolic resin is blended with
UFR. Normally, the products with highest potential for formaldehyde
emission are those bonded with UFR. During the past year,
approximately 300,000 metric tons of UFR have been used for panel
manufacturing in the U.S.

Particleboard contains between 6-8 wt% UFR (2,3). In 1984 the
annual production of UF-bonded particleboard was 5.5 million cubic
meters (3.1 billion square feet on a 3/4 inch base) in the U.S.
alone. 70% of this board was used in furniture, cabinet construction
and fixtures; 20% was used in conventional home construction, and 10%
in the manufacture of mobile homes. According to industry
sources(4), in the fall of 1985 90% of the total annual production
was capable of meeting the 0.3 ppm air chamber limit set by HUD for
manufactured housing stock (5). The production of UF-bonded
particleboard involved 48 plants in the US. Only two plants made
phenolic particleboard and only one plant produced isocyanate bonded
particleboard.

MDF contains 8-10 wt% UFR. In 1984 1.1 million m^3 was produced
in the U.S. in a total of 12 plants; 90% of this type of board is
used for furniture and cabinet work. This product is more expensive
than particleboard, but its advantage is that its edges are smooth
and dense, and thus are better capable of holding screws and hinges,
and this product need not be further treated or finished after
manufacture.

Hardwood plywood is used for interior applications only. It
contains 2.5 wt% UFR. One of the problems with plywood is that the
resin cannot be rapidly cured or dried during manufacture, because
this type of product tends to warp if moisture is unevenly removed.

As indicated above, waferboard and softwood plywood are made with
phenolic resins that are moisture resistant and do not release
significant quantities of formaldehyde if properly manufactured.

Melamin resin bonded adhesives are not yet widely used in North
America, mainly due to their cost. In Europe, they have long been
popular for making high quality interior-grade plywood. These
products emit more formaldehyde than phenolic resins, but
significantly less than UFR.

Urea-formaldehyde resins, UFR

Formaldehyde release from UF-bonded wood products depends on the
resin formulation and on curing conditions. The basic chemistry of

modern UFR manufacture and curing is deceptively simple and is not very different from that of the original invention (2). The principle is based on the condensation reaction of urea with formaldehyde in aqueous solution yielding methylol ureas that can further condense yielding methylene and ether bridged polymers:

$$HO-CH_2-OH + NH_2-CO-NH_2 = NH_2-CO-NH-CH_2-OH + H_2O \qquad (1)$$

$$R-NH_2 + R-NH-CH_2-OH = R-NH-CH_2-NH-R + H_2O \qquad (2)$$

$$2 \; R-NH-CH_2-OH = R-NH-CH_2-O-CH_2-NH-R + H_2O \qquad (3)$$

When UFR was patented in 1920 by Hanns John, Magister of Pharmacy of Prague, he clearly foresaw the unusual potential of his new materials, but the raw materials for his products were more expensive than phenolic resins at that time. The contemplated uses of these "brilliant, colorless" materials included the manufacture of window glass for automobiles and hot houses, but the inventor envisioned mainly solid, cast products, rather than wood adhesives that must be capable of forming very thin layers over large, uneven surfaces that are in constant contact with wood moisture. Today, most of the UFR production is used in manufacturing particleboard, a product that was developed during World War II in Europe in response to shortages of whole wood (6).

The main difference between early resins and the modern wood adhesives is quality control during manufacture and molar ratio of the reagents. Until very recently, most manufacturers simply mixed reagents in a given ratio for a given period of time and followed the viscosity of the resulting resin as an indication of its degree of polymerization. Today, many manufacturers follow resin synthesis with modern sophisticated analytical tools such as 13C-NMR that make it possible to analyze the actual composition of the intermediates during synthesis. Originally, UFR contained molar ratios of about F:U = 2 corresponding to the molar ratio of chemically reactive groups present in urea and formaldehyde. This molar ratio provided for sufficient formaldehyde for crosslinkage of all primary and most secondary amino groups. Even five years ago, most UFR marketed as wood adhesive resin still contained a molar ratio of F:U = 1.8, even though it was recognized that lowering the over-all molar ratio reduced the potential for post-manufacture formaldehyde release. The problem with low molar ratio resins was that they contained unreacted secondary and even primary amine groups that made the product hygroscopic. During the last ten years a tremendous amount of progress has been made in formulating low molar ratio resins and in capping unreacted methylol groups (7). Todays' adhesive resins are manufactured in three or more steps. The original step still involves large formaldehyde excess, often F:U = 4, and often involves the use of urea-formaldehyde concentrate that is made by adding urea to a concentrated formaldehyde solution. This step produces a mixture of monomethylol, dimethylol and trimethylol compounds:

$$NH_2-CO-NH_2 + 2 \; HO-CH_2-OH = HO-CH_2-NH-CO-NH-CH_2-OH \qquad (4)$$

$$NH_2-CO-NH-CH_2-OH + HO-CH_2-OH = HO-CH_2-NH-CO-N-(CH_2-OH)_2 \qquad (5)$$

Such solutions can contain up to 60 wt% formaldehyde in liquid form, while the solubility of formaldehyde in aqueous solutions is only 37 wt%. Modern resins are modified by second and third addition of urea to bring the over-all molar ratio sufficiently down to retain unreacted amino groups capable of acting as scavengers of formaldehyde that may remain unreacted or may be released by hydrolysis of unreacted methylol functions (8). In some processes additional urea is added separately to the wood furnish before drying and resination (9).

The curing conditions are equally important for reducing formaldehyde emission. The curing process is not yet fully understood. In fact, there is even still some question about the nature of the reactive resin. The latter subject is described in a later chapter by Johns. Appropriate resin cure conditions must take into account the wood moisture content and wood acidity, as well as resin concentration, temperature gradients, and press duration. In excessively cured UF bonded wood products, and in products that are stacked while still hot from the press, UFR can hydrolyse so strongly that particleboard loses internal bond strength.

Formaldehyde Complaints and Air Concentrations

Most complaints related to formaldehyde are due to defective products or improper product use. Formaldehyde is an important industrial chemical. It is extensively used in the textile industry and is present in no-wrinkle, ready-wear fabrics and a large number of consumer products and even in biological systems and living cells. Formaldehyde emitting products are the subject of a separate chapter and are listed in other publications (10). Whole wood, by itself, contains and emits only traces of formaldehyde, even though the hot pressing of forest products may cause partial hydrolysis of hemi-cellulose yielding sugars (11,12).

The problem of formaldehyde complaints is tied to the presence of formaldehyde, and is not intrinsic in aminoresins by themselves. Fully cured UF resins are odor free because they do not contain free formaldehyde. Accordingly, UF foam powder has been successfully used as a surgical wound dressing without causing irritation (13). However, the vapor pressure of formaldehyde in commercial formaldehyde, sold as 37 wt% aqueous solution, or as solid para-formaldehyde or UF concentrate, exceeds 1 Torr (14). Since the absolute threshold (15) of the pungent formaldehyde odor is 0.05 ppm, many people notice, and some are sensitive to, the presence of products that emit residual formaldehyde.

Formaldehyde emission from UF-bonded wood products has been recognized since the invention of particleboard by Fahrni (6) in 1943. Wittmann (16) recognized in 1962 that in extreme cases formaldehyde indoor air levels could reach occupational threshold levels, that these levels were increased by high load factors, temperature and humidity, and could be reduced by increasing press time and temperature, using appropriate catalysts, ammonia addition or addition of urea as a scavenger. He also showed that formaldehyde emission decreases with product age. His data indicates an initial half life of about 60 days for the products that were marketed at that time. Plath (17), Stoeger (18), Verbestel (19), Zartl (20), Neusser (21,22), Cherubim (23) and others gradually confirmed, mostly empirically, the emission charactersitics of UF-bonded wood products. In 1974, Japan introduced the first formaldehyde material emission test method, the 24-hr desiccator. This test is still in use, and is the basis for the 2 hr desiccator test that has been adopted as a standard method in the U.S. In 1977, Nestler (24) reviewed literature in the field, and later Roffael (25) and Meyer (3) published books dedicated to the subject of formaldehyde release.

When particleboard was first introduced, the risk of consumer exposure to formaldehyde emission was comparatively small as long as only moderate quantities of products were used in consumer applications. This situation changed when particleboard became popular and when its production reached millions of tons per year. This popularity caused different types of formaldehydic products, such a wall panelling, flooring, tables, cabinet work and furniture to accumulate in homes and offices, yielding load ratios of I m^2 of product surface area per 1 m^3 indoor air space.

Today formaldehyde complaints are usually due to a combination of several adverse factors involving poorly manufactured products, improper product use, and use of large quantities of new products in small, poorly or unevenly ventilated rooms. The resulting complaints can only be avoided by quality control and education at every step of use. Industrial formaldehyde levels are almost completely under control. During its use formaldehyde and its derivatives are encountered by six distinct groups of users:

Formaldehyde Manufacturer
UF Adhesive Manufacturer
Wood Product Manufacturer, Plywood, Particleboard
Architect, Home Builder, Furniture and Cabinet Maker
Indoor Air
Consumer

Each step influences the delivery and target of formaldehyde throughout the entire chain of further users. Under normal conditions, industrial handling of formaldehyde does not pose problems in the chemical factory of the basic chemical producer or the resin manufacturer, since the handling of toxic chemicals is a well established art. The acute toxic effects of formaldehyde are reasonably well known, and most countries have established occupational safety limits of about 1 ppm. In the U.S. levels are currently under revision and the subject of an advanced notice of proposed (revised) rule making (26). However, recent government field studies have shown that, in reality, occupational formaldehyde levels are only a third or less of threshold levels, even in the textile industry, the forest products industry and in pathology labs and mortuaries where concentrated formaldehyde solution is used (27,28). Typical levels and regulations are the subject of a separate chapter.

The most common human response to formaldehyde vapor is eye blinking, eye irritation, and respiratory discomfort, along with registration of the pungent odor (10). The threshold for registration of formaldehyde strongly differs among people, and its impact depends on many factors. Thus, some poeple become accustomed to what they may consider the natural odor of "wood", while others become increasingly sensitized (29). The absolute odor threshold is

0.05 ppm (15). The dose-response curve for formaldehyde odor
perception among healthy young adults ranges from 10% at 0.1 to 99%
at 1.0 ppm. Results from recent formaldehyde indoor studies confirm
the observations by Wittmann in 1962 (16) and show that formaldehyde
threshold levels for individual perception are still approached in
many living situations, and are sometimes exceeded in manufactured
housing (30,31), and in other cases of high product load
concentrations, especially in warm climates.

The incidences of perceptible formaldehyde in schools, homes, and
offices can cause uncertainty among building users about the safety
of living with formaldehyde. This uncertainty has led to the closing
of schools in Germany, Switzerland and Eastern countries. In North
America it was enhanced by the large scale installation of urea
formaldehyde foam insulation (UFFI), because a substantial part of
this material was made from small scale resin batches prepared under
questionable quality control conditions and was installed by
unskilled operators (32), often in unsuitable locations.

Several countries and agencies have responded to this uncertainty
by setting indoor air formaldehyde limits. These limits are usually
arrived at by modifying the occupational threshold levels by a factor
of ten. This factor is due to the increase in exposure time when
going from a 40 hr workplace to a home where one might spend a full
168 hr week, and by adding a safety factor of about 3 for protecting
especially sensitive individuals, such as children, old people, and
people with pre-existing sensitivities who could avoid a job
involving formaldehyde exposure but cannot avoid living in their
homes. This subject is discussed further in the chapter on
regulation.

Formaldehyde Emission Measurement and Exposure Modeling

Once the source of the emission is known and once the chemistry of
the process is established, the mass flow of formaldehyde and the
exposure level can be predicted if the appropriate parameters are
known. From a chemical viewpoint the need for free formaldehyde
ceases to exist after the pressed wood manufacture, i.e. when the UF
resin is fully cured. Thus, the presence of formaldehyde beyond the
hot press has no chemical justification and, since the advent of
recent technical improvements in every step of the manufacturing
process, it is mainly a question of quality control (10,33).
However, it is difficult and expensive to fully reduce the presence
of residual formaldehyde to the desirable trace levels for two
reasons.   Both are related to the fact that at room temperature and
50% RH wood contains 9.2 wt% moisture (34): First, moisture retains
formaldehyde quantitatively in form of methyleneglycol, and second,
wood moisture may cause slow hydrolysis of methylol end groups of the
UF polymer (3). Unfortunately, the nature of latent residual
formaldehyde is not yet fully understood. Part of it is likely in a
loosely bound state in wood moisture as methyleneglycol. Part of it
is in form of terminal methylol groups in the cured UF-resin. Thus,
the emission from wood product depends on several different factors,
including the nature of the resin, the nature of the wood, the nature
and porosity of the product, the press time, press temperature,
moisture content of the wood before and after pressing, and many

other factors (3,16-25). The literature in this field is large and
has been repeatedly reviewed.

However, on an empirical basis, the range of potential emission
behavior is reasonably well known, and the correlation between
emission measurements on product samples under standard conditions
can now be related well to the expected range of indoor air levels
under various user conditions. This subject is discussed in two
separate chapters. Thus, quality control depends on formaldehyde
emission measurements. This can be done by determination of the
formaldehyde content of the finished product, or by measuring air
levels around the product.

Formaldehyde Air Measurements. During recent years several new
measurement methods have become available. The most thoroughly
validated air measurement method is still the NIOSH chromotropic acid
test (10). In this test air is bubbled through water at a rate of l
L/min for an hour, and the formaldehyde content is then determined by
colorometric evaluation. In Europe and Japan, the acetyl-acetone
test is equally popular (3). These tests are excellent for
laboratory use, but for long-term field measurements they are awkward
and expensive. Recently, a DNPH-treated cartridge absorber (35) has
become available that makes it possible to measure air levels in the
field without liquids, tubes and beakers. Also, during the past few
years several passive samplers have become available. A sulfite-
impregnated glass paper disk in a simple diffuser tube (36) has
proven very useful and reliable in field tests in over 100,000 homes
in Canada and the U.S., but this method is not very sensitive. Very
recently, a far more sensitive passive sampler using a liquid
absorber containing 3-methyl-2-benzothiazolone hydrazone
hydrochloride (MBTH) has become available that can be used both as an
occupational personal badge sampler and as an area sampler in indoor
locations that have low levels in the 10 ppb range (37). This agent
must be developed in the field as the color dye is not indefinitely
stable.

Product tests. Clearly, the best product test is full-scale testing
of finished panels under actual use conditions. This has been done
(27,38) but is expensive, because several full-sized panels of each
product must be pre-conditioned at constant temperature and humidity
for at least a week. The next best approach is to test product
samples in air chambers under standardized conditions. A summary of
such methods is contained in Table I. A very large effort has been
made over the last three decades world-wide to develop quick,
reliable and meaningful product tests. Wittmann (16), Zartl (20),
Plath (17), Verbestel (19), Neusser (21,22), Roffael (25), HUD (5),
the U.S. Forest Products Industry (39,40), many standardization
organizations (41-43) and others have published many viable methods,
but the testing involves a combination of complex factors and there
is simply no single test that fulfills everybody's specific needs.
Table I list some of the currently accepted test methods for
formaldehyde emission from particleboard, plywood and medium density
fiberboard.

Each country has tried to find the compromise that fits its own
conditions and needs best. U.S. industry produces large quantities
of construction panels and thus needs large air chambers for testing

Table I. Formaldehyde Emission Test Methods

Class	Chamber Test	Production Test	Reference
Belgium		Perforator Value[a]:	43
Class 1		14	
Class 2		28	
Class 3		42	
Danish	0.225 m^3 chamber[b]:	Perforator Value[a]:	27,43
E-15	0.15		
P-25U		average value: 25	
P-25B	0.30	max. 10	
Finland	0.12 m^3 chamber:	Perforator[a]:	27,43
		40	
France		50	27,43
Holland		10 av.; 12 ceiling	27
Japan		24-hr dessicator[c]:	43
Norway		Perforator[a]: 30	43
Swedish	1 m^3 chamber	40	27,43
Spain		50	43
Switzerland		20	27
United Kingdom		50 average	27,43
United States			
Mobile Home:	FTM-2 Chamber[e]:	FTM-1,2hr dessicator[f]	5,41,43
	1,000-1,200 cft		
Plywood	0.2		5
Particleboard	0.3		5
MDF	0.3[g]		44
West Germany	39 m^3-chamber[h]	Perforator Test[a]:	34.43
E-1	0.12 mg/m^3	10	
E-2	0.12 - 1.2	10 - 30	
E-3	1.2 - 2.75	30 - 60	

[a]: Perforator Test: CEN-Standard EN 120-1982, (43)
[b]: Danish Air Chamber: Load: 2.25 m^{-1}; 23°C; 45% RH; 0.50 ach (currently still 0.25 ach), (27)
[c]: Finnish Chamber: Load: 1 m^{-1}, 20°C, 65% RH, 0.5 ach, (27)
[c]: Japanese Industrial Standard, JIS-A5908-1977, (10)
[d]: Swedish Air Chamber; CEN Situation Report-1983, (44): Load: 1; 23°C; 50% RH; 0.5 ach, (27)
[e]: HUD Air Chamber, FTM-2: Load 1.1; 77°F; 50% RH; 0.5 ach, (5,43)
[f]: NPA-HPMA-FI, FTM-1, 2 hr desiccator test, (42)
[g]: Industry Standard, (44)
[h]: ETH Standard Chamber: Load: 1; 23°C; 45% RH; 1 ach, (46)

these bulky products, while Denmark exports large quantities of furniture that contain small pieces and panels and thus can rely on smaller scale sampling.

However, all industries need a rapid small-scale laboratory test method for contiuous quality control of products, because such control must be conducted during the manufacturing process before

large inventories are built up and before products are sold or
shipped. In Europe, the most widely used test method is a CEN
standard method (41), the FESYP perforator test method developed in
the middle 1960s by Verbestel (19). However, this method is no
longer sensitive enough to differentiate among the products in the
lowest emission classes, such as German Class E-1, because it is
excessively sensitive to moisture content of the wood and its
findings depend on whether formaldehyde is determined
colorimetrically or by standard iodine titration (47). This test is
based on the assumption that vaporizable formaldehyde is fully
removed from small samples if they are boiled in toluene for 4 hours
at 110°C. This assumption, while never theoretically confirmed, and
strongly contested by work reported by Romeis in another chapter,
has proven a useful basis for correlation between laboratory tests
and actual air levels for individual products; but as a later chapter
in this book explains, this test is unable to provide absolute
product comparisons. In 1974, Japan introduced a 24-hr closed-jar
method (10,43) that is similar to a textile test (46), except that it
is conducted at room temperature. In the United States industry has
adopted two less sensitive 2-hr versions of the Japanese test. One
has been extensively tested by HUD in round-robin testing and
proposed as a standrad method (5); the newer version employs sealed
edges (39). In West Germany the FESYP gas analysis is also still
popular (47), even though it is now widely recognized that the
emission at the test temperature of 60°C may seriously distort
ranking of products made with different wood species or adhesives.
Another convenient method is the WKI test developed by Roffael (25),
but it also uses elevated temperatures that might distort product
rankings. However, the correlation between these quality control
methods and the air chamber tests has been well established and is
clealry sufficient for complaint investigations.

Emission Modeling

Recent work by Black, reported in a separate chapter, Mølhave (47),
and by others (48) has shown that it is now possible to quite
reliably correlate production tests to product performance if the use
conditions are well known. Indoor formaldehyde levels are determined
by the following factors:

> Formaldehyde emission rate of product
> Product surface finish
> Product use
> Temperature
> Humidity
> Load factor
> Ventilation rate
> Age

The formaldehyde emission rate has been discussed in the preceding
section. The product finish has a substantial influence on emission,
as shown in the section below.
Product Design Guidelines: Product use is a widely neglected factor.
Since UF-bonded products have essentially all the advantages of whole

wood, but are less expensive and do not crack or wharp, users have a
tendency to use them indisciminately, without regard to potential
drawbacks of the resin that can hydrolyze if it is continuously
exposed to moisture. Thus, UF-bonded particleboard is used in
roofing, for window sills, flooring and other applications where it
is only suitable if it is designed so that it is free of moisture
accumulations (10). The resulting problems could be avoided if
architects and engineers would have available a set of design
guidelines for each product that is marketed.

Environemntal Factors: The effect of temperature and humidity has
been well established (49-51):

$$C = C_o \ (1 + A[dRH]) \ \exp \ [9799(1/T - 1/T_o]   \qquad (5)$$

Since moisture equilibration, i.e. "conditioning" of wood is a slow
process that may require a week or longer depending on product
thickness, and since temperature adaptation lags by at least an hour,
the emission from wood products is not always at equilibrium. This
fact has caused non-technical people to incorrectly distrust product
performance. However, it has been found that the emission directly
reflects the daily temperature cycles of outside walls (52). Thus,
in a typical mobile home placed in a warm climate, indoor air
formaldehyde levels may change by a factor of 6 or more during a
single day. This is shown in Figure 1.
     The effect of ventilation depends on product load (53). This
subject is explained in a separate chapter:

$$C = C_o \ [KL/(N+K)]   \qquad (6)$$

where K is the porosity of the material, N the ventialtion rate in
ach, and L the load expressed in m^2/m^3. Typical curves are shown in
Figure 2. This figure shows two facts: One is that at low
ventilation rate, a small change in ventilation can bring about a
great reduction in formaldehyde level, and second, once the
ventilation rate is at 0.5 ach or above, increasing ventilation rate
does little to reduce formaldehyde levels. A typical example is
shown in Figure 3 for school furniture (54). Schools have caused
extensive problems in Europe, because they contain an accumulation of
wood products, and because they are not ventilated during the several
seasonal vacation periods. Furthermore, children have higher
metabolism than adults (10), and thus breathe relatively higher air
volumes, leading to larger pollutant doses.
     Another strong factor is age. Inasmuch as formaldehyde emission
is due to the diffusion of residual material from the center core,
the emission is proportional to the concentration, and decreases as
the concentration decreases. If all formaldehyde were present as
formaldehyde gas, or methyleneglycol, the emission process should be
strictly exponential. It has indeed been proposed that one can model
emission according to:

$$C = C_o \ \exp \ L/P \ (0.01 - C_o) \ t   \qquad (6)$$

where C_o is the starting concentration in an unventilated chamber, L
the air exchange rate per day, P the total amount of formaldehyde in

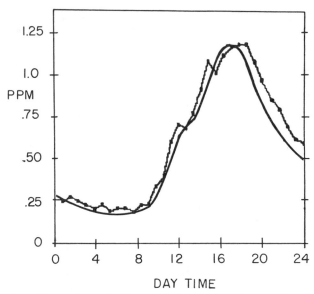

DAY TIME

Figure 1. Diurnal variations of formaldehyde air levels in a mobile home. Solid curve is calculated from product emission data; dotted curve is observed (33).

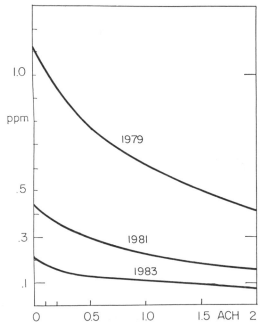

Figure 2. Formaldehyde levels as a function of ventilation rate in mobile homes containing UF-bonded wood products sold in 1979, 1981 and 1983 (33).

ppm and t the time in days. This equation is based on experience in
Swedish homes with high load factors (55). It shows that the age
effect is ventilation and load dependent. In practice, we find that
the decay follows this equation reasonably well. However, under more
exacting controlled research conditions it is observed that the decay
is not a simple exponential, but a composite, with the first decay
usually having a half life of about 60 days, while the second decay
constant depends on various manufacturing and product factors and is
about 300+30 days. Decay curves for MDF made with several different
UF adhesives (56) are shown in Figure 4. This figure shows the
correlation between F:U ratio in the resin, as well as the slopes of
the decays. As a result of this, the formaldehyde decay is very much
quicker in the first months of use, than during later use periods.
This fact was taken into consideration when the State of Wisconsin
established its formaldehyde indoor standards (10).

Emission Control and Reduction

As indicated above, formaldehyde emission depends on quality control
and on synergism between all manufacturers and users of the product.
As widely documented, properly used UF-resins with molar ratios of
F:U = 1.15 or lower are now capable of producing products that emit
only negligible formaldehyde levels under proper product use (35).
Likewise, current forest product manufacturing technology makes it
possible to produce low emitting materials by control of press
temperature, wood humdity, press duration, adhesive selection and
addition of scavengers, especially urea. One successful method for
reducing formaldehyde emission is factory treatment of fresh board

Table II. Effect of Surface Treatment

Board Conditions		Test Value (mg/m^3)
19 mm board, standard UF-adhesive		5.77
2x 120 g/m^2	edges sealed;	
	acid varnish	1.7 - 2.8
	acrylic varnish	0.56
19 mm board, reference UF-adhesive		
no finish,	all eges sealed	0.85
no finish,	edges not sealed	0.89
16 mm board, no coating,	edges not sealed	0.86
80g/m^2 melamin,	all edges sealed	0.03
	3 edges sealed	0.21
	2 edges sealed	0.35
	1 edges sealed	0.47
	not sealed	0.53
both sides melamin coated	all edges sealed	0.13
one side melamin coated, one side UF-paper		0.62
one side UF-paper ready-to-paint 0.95		

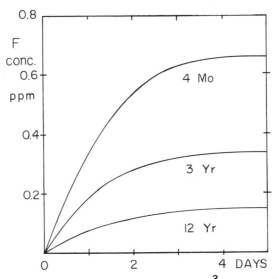

Figure 3. Formaldehyde concentration in 1 m^3 air chamber containing school chair made from plywood and solid wood with 1 m^2 surface (80% painted). Age of furniture is 4 months, 3 years, and 12 years (54).

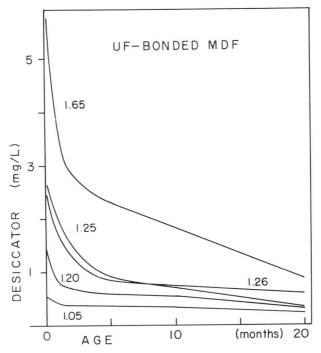

Figure 4. Formaldehyde emission of MDF as a function of age and molar ratio F:U. (27).

with ammonia (57) vapor or ammonium salts. In special applications,
where emission must be further reduced, forest products can be coated
or sealed to eliminate formaldehyde. The size of this effect (58) is
shown in Table II.

However, it has been recently shown that drilling of holes, and
decorative grooves can negate after-treatment and double emission.
This subject is discussed further in other chapters in this book.

Summary

Progress in quality control and in basic understanding of the
physical and chemical factors affecting formaldehyde emission
processes have made it possible to predict formaldehyde indoor air
levels for most use conditions. Progress in manufacturing techniques
and implementation of new technology have reduced formaldehyde
emission so much that UF-bonded wood products can now be used in
almost all applications without indoor air concentrations exceeding
0.1 pp.

Literature Cited

1.   Deppe, H. J. Holz Zentralbl. 1981, 10, 123-130.
2.   Meyer, B. "Urea-Formaldehyde Resins"; Addison-Wesley
     Publishers: Reading, MA, 1979.
3.   Maloney, T. M. "Modern Particleboard Manufacturing"; Freeman:
     San Francisco, CA, 1977.
4.   Birner, B. Wood and Wood Products 1985, 90(5), 92.
5.   "Manufactured Home Construction and Safety Standard, Air
     Chamber Test Method for Certification and Qualification of
     Formaldehyde Emission Levels," U.S. Code of Federal
     Regulations, Vol. 24, Part 3280.406, (U.S. Department of
     Housing and Urban Development), and Federal Register, Vol. 48,
     pg 37136-37195, 1983.
6.   Fahrni, F. French Patent 926,024, 1947.
7.   Mayer, J. In "Spanplatten - Heute und Morgen"; Weinbrenner,
     Ed.; DRW Verlag: Stuttgart, 1978, p. 102.
8.   Roffael, E. Holz-Zentralblatt 1973, 99(57), 845.
9.   Mansson, B, Sundin, B; K. Sirenius, Swedish Patent 434,931,
     1984.
10. Meyer, B. "Indoor Air Quality"; Addison-Wesley Publishers:
     Reading, MA, 1984.
11. Poblete, H.; Roffael, E. Holz Roh- Werkstoff 1985, 43, 57-62.
12. Stevens, M.; Schalk, J.; van Raemdonck, J. Int. J. Wood
     Preservation 1979, 1(2), 57-68.
13. Baumann, H.; Schmidt, G. Fortschritte der Medizin 1958, 76(2),
     59.
14. Meyer, B., "Urea-Formaldehyde Resins"; Addison Wesley
     Publishers: Reading, MA, 1979, pg.31.
15. Berglund, B.; Berglund, U.; Johansson, I.; Lindvall, T. Indoor
     Air. Proc. Third Int. Symp. Indoor Air Quality and Climate,
     Vol. 3, Swedish Council for Building Research: Stockholm, 1984,
     pp. 89-96.
16. Wittmann, O. Holz Roh- Werkstoff 1962, 20, 221-224.
17. Plath, L. Holz Roh- Werkstoff, 1966, 24, 312.
18. Stoeger, G. Holzforsch. Holzverwertung 1965, 17, 93-98.

19. Verbestel, J. L. FESYP Symposium, 1979, DRW Publishers:
     Stuttgart, p. 381-389..
20. Zartl, J. Holzforsch. Holzverwertung 1966, 18, 22-23.
     68, 20, 131-132.
21. Neusser, H.; Schall, W. Holzforsch. Holzverwertung 1970, 22,
     116-120.
22. Neusser, H.; Zentner, M. Holzforsch. Holzverwertung 1968, 20,
     101-112.
23. Cherubim, M. Holz Roh- Werkstoff 1976, 34, 449-452.
24. Nestler, F. H. M. "The Formaldehyde Problem in Wood Based
     Products-An Annotated Bibliography," FPL Report FPL-8 U.S.
     Department of Agriculture, 1977.
25. Roffael, E. "Formaldehydabgabe von Spanplatten und anderen
     Werkstoffen"; DRW Publishers: Stuttgart, 1982.
26. "Occupational Exposure to Formaldehyde," Occupational Safety
     and Health Agency, U.S. Department of Labor, Federal Register
     1985, 50, 15179 and 1985, 50, 50412-50499.
27. Sundin, B. Proc. Int. Particelboard Symposium, 1985, 19, 200.
28   "Formaldehyde: Determination of Significant Risk," U.S.
     Environmental Protection Agency, Federal Register 1984, 49,
     21870.
29. Ulsamer, A. G.; Beall, J. R.; Kang, H. K.; Frazier, J. A.
     Hazard Assessment of Chemicals 1984, 3, 337.
30. Hanrahan, L. P.; Dally, K. A.; Anderson, H. A.; Kanarek, M. S.;
     Rankin, J. Am. J. Public Health 1984, 74, 1026-1027, and J.
     Air Pollution Control Assocation 1985, 35(11), 1164.
31. Stock, T. H.; Monsen, R. M.; Sterling, D. A.; Norsted, S. W.
     78th Annual Meeting Air Pollution Control Assoc., Air Pollution
     Control Assoc.: Detroit, 1985.
32. "Ban of Urea-Formaldehyde Foam Insulation," Consumer Product
     Safety Commission, Federal Register, 1982, 47, 14366.
33. Meyer, B.; Hermanns, K. J. Air Pollution Control Assoc. 1985,
     35, 816-821.
34. Skaar, C. "Wood-Water Relationships," Adv. Chem. 1984, 207,
     127-172.
35. Lipori, F.; Swarin, S. J. Environmental Science and Technology,
     1985, 19, 70-74
36. Geisling, K. L.; Miksch, R.R.; Rappaport, S.M. Anal. Chem.
     1982, 82, 140.
37. Miksch, R. R., personal Communication, 1985.
38. Korff, C, Center for Surface Technology, Haarlem, Holland,
     unpublished data, 1985.
39. "Small Scale Test Method for Determining Formaldehyde Emission
     from Wood Products, Two Hour Dessicator Test, FTM-1," National
     Paricleboard Association, Hardwood Plywood Manufacturers
     Association, Formaldehyde Institute and U.S. Department of
     Housing and Urban Development, Federal Register, 1982, 48,
     37169.
40. "Large Scale Test Method for Determining Formaldehyde Emission
     from Wood Products; Air Chamber Method, FTM-2" National
     Particleboard Assocaiton, Hardwood Plywood Association, U.S.
     Department of Housing and Urban Development, Federal Register,
     1982, 48, 37169.

41. "Particleboard-Determination of Formaldehyde Content-Extraction Method Called Perforator Method," European Standard EN-120-1982, European Committee for Standardization, Brussels, 1982.
42. "Particleboard-Determination of Formaldehyde Emission under Specified Conditions; Method Called: Formaldehyde Emission Method," European Standard Situation Report EN-N76E-1983, European Committee for Standardization, Brussels, 1983
43. "Materials and Fittings, A-5906-1983 Medium Density Fiberboard; A-5907-1983 Hard Fiberboards, A-5908-1983 Particleboard, A-5909-1983 Dressed Particleboard, A-5910 Dressed Hard Fiberboard," Japanese Industrial Standards, (Official English Translation, available through the American National Standard Institute, New York), 1985
44. "LOFT paneling and Mobile Home Decking," and "Fiberwood .3" Weyerhaeuser Corporation, Tacoma, WA, 1981 and 1984.
45. Sundin, B. personal communication, 1985.
46. "AATCC Test Method 112-1975," American Association Textile Chem. Colorists, AATCC Technical Manual, 1975, Vol. 55.
47 Mølhave, L.; Bisgaard, P.; Dueholm, S. Atmospheric Environment, 1983, 17, 2105-2108.
48. Matthews, T. G.; Reed, T. J.; Tromberg, B. J.; Fung, K. W.; Thompson, C. V.; Hawthorne, A. R. "Modeling and Testing of Formaldehyde Emission Characteristics of Pressed-Wood Products," Consumer Product Safety Commission, CPSC-IAG-84-1103, 1984.
49. Berge, A.; Mellegaard, B.; Hanetho, P.; Ormstad, E.B. Holz Roh-Werkstoff, 1980, 38, 251.
50. Myers, G. E. Forest Products Journal, 1985
51 Godish, T, J. Air Pollution Control Association, 1985, 35(11), 1186.
52. Meyer, B.; Hermanns, K. J. Environ. Health 1985, 48, 57-64.
53. Hoetjer, J. J; Koerts, F. Holz Roh-Werkstoff, 1981, 39, 391.
54. Marutzky, R. "Formaldehydprobleme bei der Be- und Verarbeitung von Holz und Holzwerkstoffen"; Fraunhofer-Institut für Holzforschung, Wilhelm-Klauditz-Institut: Braunschweig, Germany, 1985.
55. Sundin, B. 3rd Medical Legal Symp. Formaldehyde Issues, Professional Consultants in Occupational Health: Maryland, 1982.
56. Meyer, B.; Hermanns, K. J. Adhesion, 1985, 17, 297-308.
57. Smedberg, O.; Larsson, K. U.S. Patent 4,255,102, 1981.
58. Dueholm, S. Saertryk af Limspecialisten 1985, A/S Kemi Casco: Stockholm, p. 1-5.
59. Rybicky, J. Wood Fiber Sci. 1985, 17(1), 29-35.

RECEIVED January 14, 1986

Formaldehyde Emissions: Hardwood Plywood and Certain Wood-Based Panel Products

William J. Groah

Hardwood Plywood Manufacturers Association, 1825 Michael Faraday Drive, Reston, VA 22090

Hardwood plywood products are decorative in nature
and are designed for interior use. Over 95% of all
hardwood plywood is made with urea-formaldehyde
adhesives. Responding to concerns about formaldehyde and certain wood products, test methods for
measuring surface emissions were developed in the
early 1980's. Emissions from most hardwood plywood
and particleboard products have decreased 65% to 95%
in recent years primarily by use of low emitting UF
adhesives and/or scavengers. Good correlation has
been demonstrated between product test methods and
indoor levels of formaldehyde in experimental manufactured homes. Decorative surface finishes can act
to either increase or decrease surface emissions,
depending on the nature of the finish and the substrate.

Lines of demarcation between hardwood plywood, softwood plywood
and certain other wood based panel products have become less
distinct in recent years. One of the most important distinctions
in respect to formaldehyde emission potential is that softwood
plywood is typically bonded with phenol-formaldehyde while
hardwood plywood is typically bonded with urea-formaldehyde.
Phenol-formaldehyde adhesives are more stable and have less
tendency to emit formaldehyde than do urea-formaldehyde adhesives.

Some important features of hardwood plywood:

1. The face veneer is used to describe the product. Oak plywood,
   for example, will have oak face veneer; the inner layers and
   back veneer will likely be of some other product or species.
2. Most hardwood plywood products are decorative in nature.
3. Most hardwood plywoods are designed for interior application.
4. Face veneers typically are high quality. For many face

0097–6156/86/0316–0017$06.00/0
© 1986 American Chemical Society

species the cost of logs is high and faces are sliced thin,
ranging from about 1/20" to 1/100" in thickness.

5. Because face veneers are decorative and thin, a colorless glue
    line is desired to prevent discoloration on the face.

6. Urea-formaldehyde adhesives are predominate in the manufacture
    of hardwood plywood. Well over 95% of all hardwood plywood
    consumed in the U.S. is made from UF adhesives.

Apparent U.S. consumption of hardwood plywood in 1983 was
4.3 billion square feet surface measure having a value of about
1.1 billion dollars. About 2/3'rds, on a surface measure basis,
was imported, Indonesia being the primary exporting country, with
Korea, Taiwan, Philippines and Malaysia also being important
factors.

Formaldehyde emission and/or formaldehyde space level
potential can be related to both construction type and product end
use. While the American National Standard for Hardwood and
Decorative Plywood (1) references eight different types of construction, three are most important in the context of formaldehyde:

Veneer core - 3, 5, 7, 9 ply and greater

Particleboard core - 3 ply

Medium Density Fiberboard core - 3 ply

Both particleboard and MDF core are characteristically 3-ply
and have two potential sources of formaldehyde: the adhesive used
to adhere the hardwood face and back to the core, and the adhesive
binder used in the manufacture of the particleboard or MDF.
Hardwood plywood manufacturers are typically not vertically
integrated and do not produce composition board cores, thus are
dependent on other companies or plants for particleboard and MDF.

The single largest end use for hardwood plywood is interior
wall panels, generally 3-ply and 1/4" and thinner, and frequently
machined with decorative v-grooves. Furniture, cabinets, door
skins and a number of specialties complete an array of end use
products. Many of the non wall panel products can be characterized as being industrial panels and are of 5 or more ply veneer
core, 3-ply particleboard core, or 3-ply medium density fiberboard
(MDF) core construction. Broad end use patterns indicate that
interior wall panels represent approximately 55% of total hardwood
plywood consumption. Furniture, cabinets, and fixtures represent
about 30%, and door skins and specialty products about 15% (2).
Potential sources of formaldehyde in two of the more typical
hardwood plywood constructions are displayed in Figure 1.

Formaldehyde Issue Benchmarks

The potential for elevated ambient formaldehyde levels became
apparent in manufactured housing during the late 1970's. Federal
standards governing the construction of manufactured or mobile
homes first became effective in 1976, a period which coincided
with the dramatic increase in cost of energy and the tightening

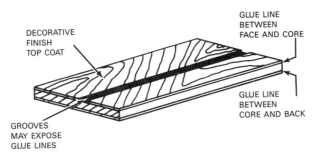

3-PLY WALL PANEL

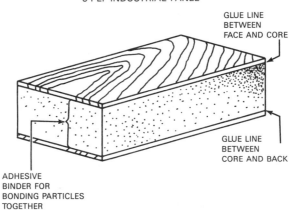

3-PLY INDUSTRIAL PANEL

Figure 1. Potential sources of formaldehyde in two typical hardwood plywood constructions.

of homes of all types. In the late 1970's, the domestic hardwood plywood industry became significantly involved in responding to concerns about hardwood plywood as a potential source of formaldehyde emissions. The primary industry focus has been on wall panel use in manufactured housing because of low air exchange rates coupled with high product usage or loading. A few years ago it was not uncommon for decorative wall paneling to be used on almost all interior walls in mobile homes. The use of wood wall panels in manufactured homes has declined in recent years but is still significant.

One of the first major efforts was to investigate how formaldehyde emissions from products could be determined. This eventually resulted in two industry developed test methods: the two hour desiccator test designated as FTM 1, and the large laboratory chamber test designated as FTM 2 (3,4). Concomitant with the assessment of analytical techniques and the development of test methods was an effort to determine the potential for reducing formaldehyde emissions from hardwood plywood. A product emission survey of hardwood plywood products was made about 1980 to determine the then current state-of-the-art. In obtaining samples for the the survey, plywood was characterized as standard, low emitting and odor free. This terminology was selected to be similar to that used in Japan and perhaps well understood by other countries in the Asian region. Products were obtained from various companies in Asia and North America. The results of the survey appeared to demonstrate that emissions could be reduced by 65% to 95%.

Reducing Formaldehyde Emissions

Technical considerations, resin cost, and resin availability have and are dictating low emitting UF systems as the primary substitute for standard UF adhesives. Relatively small quantities of hardwood plywood have been made with polyvinyl acetate and phenol formaldehyde, the two most likely substitutes. Cost is a primary disadvantage of PVA types and certain use parameters and the dark color of phenol limit that adhesive to certain hardwood plywood factories and for certain applications.

Reduction in the emission characteristics of unfinished hardwood plywood is currently being achieved primarily by the use of low formaldehyde to urea molar ratio formulations. For the manufacture of hardwood plywood and particleboard, formaldehyde to urea molar ratios have been reduced to a range of 1.15/1 to 1.3/1. An important caveat; low F/U ratios perhaps should be considered a proxy for the potential to reduce emissions through improved urea-formaldehyde adhesive technology rather than the exclusive means for improvement. Reducing the F/U ratio is not always the most effective way of reducing emissions in consideration of the variety of hardwood plywood constructions, products, and thick-nesses.

Surface applied post treatments are also commonly being used. Myers (5) has documented the effectiveness of a laboratory applied ammonia treatment and also a urea containing coating to hardwood plywood. In practice most commercial treatments are applied by roller coaters and the effectiveness of the treatment depends not

only on the treatment material but also the application rate and
the nature of the product being treated. Exact formulations are
typically proprietary but most treatments are believed to contain
some ammonia or urea compound and are applied at rates that
achieve a 30% to 85% reduction in formaldehyde emissions. Some
manufacturers use both low emitting UF adhesive systems and post
treatments.

Two Product Related Factors That Can Effect Emissions

Surface finishes can be an important factor in either increasing
or decreasing emissions. This became apparent as formaldehyde
emissions decreased as a result of changes in UF adhesives. Wall
panel products can be segmented by the type of decorative surface
finish in order of commercial importance.

> Paper Overlays - (40% of wall panel consumption) are 1 to 2
> mil printed paper films adhered to lauan plywood with PVA or
> other adhesive. Paper films are available in pre-topcoat or
> non pre-topcoat varieties.

> Printed - (35%) surfaces are decorative pattern or simulated
> wood grain effects created by the application of liquid
> applied basecoats, inks and protective topcoats to lauan or
> other tropical hardwoods.

> Natural Hardwoods - (18%) describe essentially transparent
> finish systems on species such as walnut, oak, birch, pecan,
> cherry, etc.

> Vinyl Overlay - (7%) are 2 mil or thicker printed vinyl films
> adhered to lauan plywood with PVA or other adhesives.

The domestic hardwood plywood industry has been trending
towards the use of water based topcoats for some paper overlay,
printed, and natural hardwood paneling products to reduce volatile
organic compound emissions. To achieve desired surface product
properties formaldehyde is often a component of the topcoat.
There have been efforts to reduce the amount of emittable formaldehyde in topcoats or to reformulate to eliminate formaldehyde as
a component.
     Finishes in some cases also appear to reduce emissions from
wall paneling products. The effectiveness of a vinyl film overlay
was evaluated using high emitting hardwood plywood wall panels
(6). Formaldehyde emissions from the vinyl surface of plywood
were compared with the back or unexposed plywood surface using
both the large chamber and the two hour desiccator. This comparison indicated that a 2-mil vinyl was about 90% effective in
reducing emissions.
     The number of V-grooves can be a factor, particularly when
only post surface treatments are used prior to panel grooving on
relatively high emitting panels. Matched specimens were carefully
selected for desiccator testing to compare the effect of number of
grooves from zero to 16 (one groove for each desiccator sample
surface) from a group of panels where the improvement in emission

characteristics was achieved primarily by a post surface treatment. A typical v-grooved 4' x 8' wall panel has five to seven
grooves which translated to 4 to 6 grooves in this study. The
data plot of Figure 2 suggests that a 30% to 40% increase in
emission rate could theoretically result when grooves are cut
following post treatments of panels made with standard UF
adhesives.

Manufactured Home Regulations

Particleboard decking and hardwood plywood wall panels can
represent 80% to 90% of the total exposed surface of formaldehyde
containing wood based products in new mobile homes. Kitchen
cabinets, vanities, shelving and other built-ins are primarily
made from industrial particleboard, MDF or hardwood plywood
panels.
     On February 11, 1985, a rule establishing product standards
for hardwood plywood and particleboard used in manufactured
housing became effective. The U.S. Department of Housing and
Urban Development has designated a chamber loading rate of 0.29
sq ft/cu ft and chamber level of 0.2 ppm for hardwood plywood, and
a 0.13 sq ft/cu ft loading rate, and 0.3 ppm level for particleboard. Industrial Panels, that is panels that are composite in
nature and are used for applications other than wall paneling,
also have been defined by HUD in January 1985 to be tested at a
loading rate of 0.13 sq ft/cu ft to conform to a 0.3 ppm chamber
level. HUD mandates the large chamber as the primary test method
to be used for initial product compliance and to be conducted
thereafter at a frequency of once a quarter.
     In-plant quality control and routine acceptance testing by
property verification organizations such as the Hardwood Plywood
Manufacturers Association and the National Particleboard Association require a method more efficient than the chamber for
routinely monitoring trends in emission characteristics of products. The relationship between chamber and the small scale
desiccator test observations is illustrated by a series of 76
tests accomplished during the past year on hardwood plywood wall
panel products at a chamber loading rate of 0.29 sq ft per cu ft:

$$Y = 0.62 \ X + 0.005 \qquad\qquad\qquad [1]$$

Where: Y = The chamber value
        X = The average desiccator value of all panels placed
            in the chamber

NOTE: The equation above is generic in nature and should not
necessarily be used to describe the small and large scale test
relationship for all wall panel products.

     The HUD product standards are tied to the objective of
providing a 0.4 ppm ambient target level in new manufactured
homes. The hypothesis that product emission standards can be
related to ambient formaldehyde levels was tested in a HUD sponsored project (7) that involved constructing four experimental
mobile homes and comparing home observed formaldehyde levels with

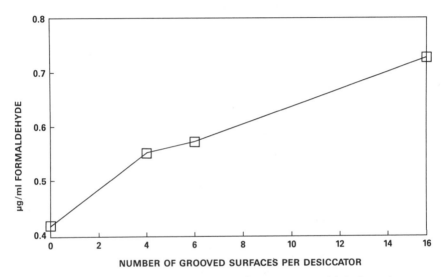

Figure 2. Relationship between desiccator formaldehyde values
and number of V-grooves in desiccator samples.

product emissions in laboratory tests including the large chamber.
This experiment demonstrated that home levels could be related
reasonably well to chamber levels when plywood paneling and
particleboard decking were tested together at manufactured home
loading rates. Moreover, it was demonstrated that chamber levels
of paneling and decking tested together (in combination) relates
to values obtained by the addition of the chamber values for
paneling and for decking when tested individually.

$$Y = 0.81X - 0.09 \qquad\qquad\qquad [2]$$

Where: Y is the chamber value for decking and paneling
        tested in combination.

        X is the decking value plus the paneling value when
        the products are tested separately.

Concluding Statements

While the initial concern for formaldehyde emissions in the
hardwood plywood industry was in the wall paneling sector there is
a strong and decided movement by many manufacturers to apply new
low emitting adhesive technology to other hardwood plywood
products. Low emitting UF products are nearing the emission
characteristics of certain other resin systems assumed to be
likely substitutes.
    The HUD rule has had an effect far beyond the products
designed for manufactured housing. Some companies that make wall
panels either do not or may not know where the product will be
used. Many companies have elected to use low emitting products
that meet HUD product standard requirements even if they know the
product will not be used in manufactured homes.
    It would appear that the wall paneling industry, on average,
has probably been able to achieve a 70% to 95% reduction in
formaldehyde emissions and still maintain the use of urea-formaldehyde adhesives.

Literature Cited

1. American National Standards Institute. American National
    Standard for Hardwood and Decorative Plywood, ANSI/HPMA HP
    1983.
2. Hardwood Plywood Manufacturers Association, Comments before
    the Environmental Protection Agency, regulatory investigation
    of formaldehyde exposures determined to be within section 4(f)
    of the toxic substances control act, 1984.
3. National Particleboard Association, Hardwood Plywood Manu-
    facturers Association, Formaldehyde Institute. October 10,
    1983. Small scale test method for determining formaldehyde
    emissions from wood products, two-hour desiccator test, FTM 1,
    Reston, VA.
4. National Particleboard Association, Hardwood Plywood Manu-
    facturers Association, October 10, 1983. Large scale test
    method for determining formaldehyde emissions from wood
    products, large chamber method, FTM 2, Reston, VA.

5. Myers, G. E., Formaldehyde dynamic air contamination by
    hardwood plywood: effects of several variables and board
    treatments. Forest Products Journal, 1982; 32(4):20-25.
6. Groah, W., G. Gramp, M. Trant, 1984. Effect of a decorative
    vinyl overlay on formaldehyde emissions. Forest Products
    Journal, 1984; 34(4):27-29.
7. Singh, J., R. Walcott, C. St. Pierre, T. Ferrel, S. Garrison,
    G. Gramp, W. Groah. "Evaluation of the relationship between
    formaldehyde emission from particleboard mobile home decking
    and hardwood plywood wall paneling as determined by product
    test methods and formaldehyde levels in experimental mobile
    homes." U.S. Dept. of Housing and Urban Development, 1982.

RECEIVED January 14, 1986

3

Formaldehyde Release from Wood Panel Products Bonded with Phenol Formaldehyde Adhesives

J. A. Emery

American Plywood Association, 7011 South 19th Street, Tacoma, WA 98466

Both the published literature and previously unpublished information obtained by the structural panel industry indicate that formaldehyde levels associated with panel products glued with phenol formaldehyde adhesives are extremely low. Large dynamic chamber tests which simulate conditions that might be found in tightly sealed residences indicate consistently that formaldehyde levels associated with freshly manufactured phenolic panel products are less than 0.1 parts per million. The data, as well as theoretical considerations, also indicate that the amount of formaldehyde contributed to the environment by phenolic panel products should rapidly approach zero as the small quantity of formaldehyde initially present in the products is released.

Virtually all wood panel products such as plywood and particleboard are manufactured using either urea formaldehyde or phenol formaldehyde adhesives. Urea formaldehyde adhesives are used in hardwood plywood and in certain types of particleboards. These adhesives are not waterproof, and products made with them are normally used indoors for paneling, furniture, shelving and floor underlayment.

Phenol formaldehyde, on the the other hand, is used to make the waterproof adhesives that are used in structural wood panel products such as softwood plywood, oriented strand board, waferboard and exterior (phenolic) particleboard. These products are commonly used for roof, floor and wall sheathings, exterior sidings, concrete forms and in pallets and numerous other products.

Although formaldehyde emissions from some products glued with urea formaldehyde adhesives can cause indoor air quality problems under certain conditions, such problems have not been associated with phenol formaldehyde-bonded (phenolic) products. Unfortunately, however the commonplace usage of the generic terms "particleboard" and "plywood" has failed to distinguish between product types and has led to a great deal of confusion among consumers.

Because phenolic panels have not presented formaldehyde-related problems in the marketplace, there has not been much need to develop information on formaldehyde emissions from these products.

Nevertheless, a considerable amount of information has been generated to satisfy curiosity and to answer inquiries concerning emissions from phenolic panel products. This information is summarized in this paper under three primary subject headings: (1) Theoretical Considerations; (2) Literature Review; (3) Previously Unpublished Information.

Theoretical Considerations

The chemical and physical characteristics of phenolic resins and adhesives made from them suggest that formaldehyde emissions should be very minor (1). One reason for predicting low emissions is that very little residual free formaldehyde is present in prepared phenolic resins. This low free formaldehyde content is due to both the use of low formaldehyde to phenol mole ratios in resin preparation and to the tendency of nearly all the formaldehyde to react irreversably with the phenol.

Another reason for predicting low emissions is that the small amount of residual formaldehyde that might be present in the prepared resin is diminished even farther by reactions which occur when the resin cures. Phenolic resins are cured under heat and pressure in a hot-press, usually under highly alkaline conditions. Curing temperatures are usually in the range of 130-220°C. Under these conditions, unreacted formaldehyde continues to react with phenol to form larger phenol formaldehyde polymers. Also, some formaldehyde reacts with various chemical constituents in the wood. Moreover, some formaldehyde is probably converted to methyl alcohol and formic acid by way of the Cannizzaro reaction (1).

A third reason for predicting very low emissions of formaldehyde from phenolic panels is that the cured resin is extremely stable and does not break down to release additional formaldehyde, even under extremely harsh environmental conditions (2). The high resistance of phenolic resins to deterioration under severe service conditions is, of course, a principal reason they are used so widely in making exterior types of wood panel products. Because of their chemical stability the U.S. Environmental Protection Agency has declared that phenol formaldehyde resins represent a "consumptive use" of formaldehyde, meaning that formaldehyde is irreversibly consumed in its reaction with phenol so that the formaldehyde loses its chemical identity (3).

Any formaldehyde that might be present initially in fresh phenolic panels, would be expected to diminish through time, since additional formaldehyde is not released from a breakdown of the resin. Thus, barring contamination from other sources, formaldehyde emissions associated with thoroughly aged phenolic panels should be nil.

Literature Review

The formaldehyde emitting potential of wood panel products can be evaluated in numerous ways, including the use of dynamic chamber tests (tests involving chambers which are ventilated and simulate real-world conditions); static (unventilated) tests, such as desiccator and "equilibrium jar" methods; and chemical extraction tests, such as the European Perforator Test. This review will emphasize the results from dynamic chamber tests, especially the "large scale" dynamic chamber test, since the results of such tests can generally be compared and since the results are more representative of real-world situations than are the results of other tests. Data from static tests and chemical extraction tests are more abstract than the results of dynamic chamber tests, and such data must therefore be correlated with some type of dynamic test in order to be useful in terms of evaluating actual potential exposures.

Nestler (4) thoroughly reviewed the worldwide literature on formaldehyde emissions from wood products published through January, 1977. According to Blomquist (1), Nestler's literature review includes only three citations which even mention phenolic adhesives, and none of these citations made specific mention of any problems associated with the use of phenolic panels.

Since Nestler's review was published, some additional information on formaldehyde emissions from phenolic panels has appeared in the literature. Information obtained using dynamic test chambers is summarized in Table I. Perforator and two-hour desiccator data are summarized in Table II.

As indicated in Table I, dynamic chamber test data have been obtained in investigations using chambers ranging in size from 0.003 m^3 (0.1 ft^3) to 28 m^3 (1000 ft^3). Besides this large chamber size variation, the studies also varied widely with respect to the temperatures, relative humidities, loading rates, and air exchange rates used for testing. Because of the wide variations among the studies with respect to these test parameters, it is not possible to make many inferences from the data presented. However, some general trends are evident, and certain relationships developed from studies involving urea formaldehyde systems make it possible to make a few generalized observations concerning the data.

Although Table I indicates that formaldehyde levels ranging from 0.01 - 0.3 parts per million (ppm) have been observed in studies using dynamic test chambers, values of 0.1 ppm or lower were observed in most of the investigations. Those studies in which the higher levels were found (i.e., the first two studies summarized in the table) used very small test chambers (0.003 m^3) and relatively high temperatures, humidities, and loading rates. Lower levels are shown for those studies wherein the test parameters approximated "real-world" conditions (large test chambers using temperatures, humidities, loading rates, and ventilating rates approximating those found in living areas).

The higher formaldehyde levels associated with the first two studies summarized in Table I can probably be attributed primarily to the relatively high temperatures employed. Numerous investigations have shown that formaldehyde levels increase exponentially with temperature (5-7). Several studies have shown that formaldehyde levels associated with wood panel products can increase by more than a factor of 3 as the temperature increases from 25°C to 40°C (8). The exponential function developed by Berge, et.al, (5) is commonly used to adjust formaldehyde data for temperature (9). If this function were applied to the data of Table I in order to adjust all formaldehyde levels to a common temperature of 25°C, the corrected levels

Table I.   Summary of Published Dynamic Chamber Test Data for Phenolic Panels

Product Type[1]	Dynamic Test Chamber Parameters					HCHO Levels (ppm)[3]	REMARKS	REFS.
	Vol. (M³)	Temp. (°F)	R.H. (%)	Loading (m²/m³)[2]	Air Changes Per. Hr.			
PB	.003	95	67	6.7	1.7-6.6	.05-.3[4]	Data represent measurements on 4 samples over 16 days before and after aging for 90 days @ 23°C, 44% R.H.: .1 - .3 ppm unaged; .05 - .1 ppm aged.	[10]
PB	.003	77-104	75	19.2	1.15	.12-.18[4]	Emissions measured for two boards at two different temperatures.	[8]
SWPW	.21	77	50	.1-10	1-4	.1[4]	Chamber value estimated from study of emission rates at .1 ppm background level (rate=0).	[18,19]
WB	28	75	50	.43	.5	.01-.04	Data represent tests on 24 samples from 8 manufacturers.	[20]
WB	28	75	50	.43	.5,1	.1,.06[4]	The 2 formaldehyde levels correspond to the 2 ventilation rates shown	[12]
SPP	28	75	50	.43	.5,1	.1,.06[4]	- " -	[12]
PB	28	75	50	.43	.5,1	.1,.06[4]	- " -	[12]

(1) PB = particleboard; SWPW = softwood plywood; WB = waferboard; SPP = southern pine plywood.
(2) Loading is expressed in terms of square meters of panel surface area per cubic meter of chamber volume.
(3) Parts per million parts of air, volume basis.
(4) Data values taken from graphs.

Table II. Summary of Published Two-Hour Desiccator
and Perforator Test Data for Phenolic Panels

Product Type(1)	Formaldehyde Levels		Remarks	Ref.
	2-Hr. Desiccator (μg/ml)(2)	Perforator (mg/100g)(3)		
PB	1.09-1.37		Values represent 4 different particle-boards.	(10)
WB		.16		(10)
WB	.11-.27	.29-.85	Two hr. desiccator values represent tests on 45 samples from 11 manufacturers. Perforator values represent 24 samples from 8 manufacturers.	(20)
PP		.6		(14)
SWPW	.18			(18,19)

(1) PB = particleboard; WB = waferboard; PP = phenolic plywood; SWPW = softwood plywood.

(2) FTM 1-1983 procedure (16) was used in all studies. Values represent micrograms of HCHO per ml distilled water. In this test 25 ml distilled water in a petri dish is placed in desiccator with eight 7 cm x 12.7 cm specimens for 2 hours; and the water is then analyzed for formaldehyde.

(3) Total HCHO extractable with boiling toluene, mg. formaldehyde per 100 grams dry wood (21).

would all be 0.1 ppm or lower. Such an adjustment would make the
emission data from the small chamber tests similar to those from the
large chamber tests summarized in the table.

It is important to note here that higher temperatures probably
increase emissions from phenolic panels simply by accelerating the
release of that small amount of residual formaldehyde that originates
from the adhesive and subsequently becomes adsorbed to the wood sub-
stance and water in the wood. Because phenolic resins are very
stable chemically, any temperature-related increase in emissions
would not be expected to be associated with resin degradation.
Consequently, temperature would be expected to exert much less
influence on emissions from panels which have been aired out than
from fresh panels. Indeed, this trend is shown by the data, as discussed below.

The information presented in Table I also indicates that the
loading and ventilation rates for those two studies in which the
higher formaldehyde levels were found (8,10) were higher than for
the other studies summarized. The influence of these factors on
formaldehyde levels has not been clearly explained, however, since
the amount of data pertaining to phenolic panels is so limited and
since the literature appears to be contradictory. Studies of urea
formaldehyde-bonded systems generally indicate that emissions
increase with higher loadings and decrease with higher ventilation
rates (6,11). Moreover, Meyers (11) has shown that there is often
a good relationship between the ratio of ventilation and loading
rates (N/L ratios) and formaldehyde concentration in controlled chamber experiments. Indeed, the data presented in two of the studies
summarized in Table I (10,12) appear to be in general agreement with
these trends, since the data show decreases in formaldehyde levels
corresponding to increased ventilation at constant loading. However,
other studies have indicated that emission levels from very low emitting products are not influenced significantly by loading or ventilation rates (6). More research on these relationships is needed.

The effect of panel age on formaldehyde release was investigated
in the first study summarized in Table I, and this variable was evidently very important with respect to the formaldehyde levels
measured. As noted in the Remarks column in the table, formaldehyde
levels ranged from 0.1 - 0.3 ppm for freshly manufactured specimens,
while levels in the range of only 0.05 - 0.1 ppm were associated with
matched specimens that had been aired out for 90 days at 23°C and
44% relative humidity. This aging effect is consistent with the
theoretical considerations discussed earlier and with test results
to be presented later in this report.

The two-hour desiccator and Perforator test results shown in
Table II are also indicative of very low formaldehyde levels for
phenolic panels. As with most of the results obtained in dynamic
chamber tests, the uniformity of these test results, both within and
between studies, indicates that the various phenolic panel products
are quite similar with respect to their emitting potential.

Twenty-four hour desiccator tests were also conducted in some of
the studies summarized in Tables I and II (8,10), but the results are
not shown since different test procedures were used in each of the
studies and the data are, therefore, not comparable.

In addition to the studies summarized in Tables I and II, Meyer
(13) measured formaldehyde emissions from samples of phenolic plywood, waferboard and particleboard using a modified version of the
Japanese Industrial Standard, which is a type of 24-hour desiccator
test. Emissions from wood veneer and a urea formaldehyde-bonded
particleboard with known emission characteristics were also measured.
This researcher used the desiccator test results, along with information from the literature, to estimate the maximum amount of formaldehyde that might potentially be contributed by phenolic panel products
to indoor air. Assuming a loading factor of $1.18-m^2/m^3$ and no ventilation, calculations showed that phenolic panels would contribute
less than 0.05 ppm. Assuming a ventilation rate of one-half air
change per hour, calculated levels were below 0.0025 ppm. The tests
also indicated that formaldehyde levels associated with wood veneer
alone (without any added adhesive) were about the same as levels associated with phenolic panels. No background formaldehyde levels
were reported, however; and considering the findings of studies which
are discussed later, background levels could easily have been as high
as those reported for both the veneer and the phenolic products.
Regardless of background level considerations, the study generally
indicates that phenolic panels emit extremely low levels of formaldehyde, thus corroborating the findings of the studies discussed
earlier.

Sundin (14) also measured formaldehyde emissions from a sample
of phenolic plywood in a static chamber (no air exchange) with a
volume of 15 m^3. The temperature was maintained at 20°C and the
loading rate was 1 m^2/m^3. Relative humidity was not controlled, but
was reported to be generally in the range of 30-50%. The exact formaldehyde level measured in the chamber was not reported, but the
author concluded that ... "the emission from phenol formaldehyde (PF)
- glued plywood is extremely low and in practice is negligible ... "
A graph is presented that indicates the level was below 0.2 ppm. As
indicated in Table II, a Perforator value of 0.6 mg/100g was also
reported for this plywood.

Roffael (15) measured formaldehyde emissions from a phenolic
particleboard using the "WKI-Method" which involves suspending small
samples over 50 cm^3 of distilled water in tightly closed polyethylene
bottles and measuring formaldehyde levels in the water after varying
times. Temperatures were maintained at 42°C. This work indicated
that formaldehyde release from the phenolic particleboards ceased
after a relatively short reaction period (approximately 96 hours).
This finding is consistent with the resin stability considerations
discussed previously under theoretical considerations.

Previously Unpublished Information

Much of the information pertaining to formaldehyde emissions from
phenolic panels has been obtained by manufacturers of these products
but has not been published previously in the open literature. This
information has been obtained primarily to form a basis for answering
consumer inquiries.

American Plywood Association Study. Probably the most extensive
study of phenolic panel emissions was conducted by W. F. Lehmann of
Weyerhaeuser Company for the American Plywood Association. In this

investigation, formaldehyde emissions from representative samples of
all major types of phenolic panels were measured using a Large-Scale
Dynamic Chamber (LSDC) and two-hour desiccator tests.

Panel types included southern pine and Douglas-fir plywood,
oriented strand board from two different manufacturers, waferboard,
and a phenolic particleboard. For each product type, five 1.2m x
2.4m (4 ft x 8 ft) panels were obtained during a single shift. The
panels were kept stacked together during shipping and storage until
three days prior to teting, when they were placed in racks in order
to allow air circulation around each panel.

Strips measuring 15.2 cm in width were cut from the centers of
four of the five panels, parallel to the shorter panel dimension.
Four 2-hour dessicator test specimens, each measuring 7 cm x 12.7 cm
(2-3/4 in. x 5 in.), were cut from each strip and conditioned overnight. Two desiccator tests were conducted for each product type,
with each desiccator containing eight specimens. Tests were performed in accordance with standard procedure FTM 1 (16).

Specimens for the LSDC tests were prepared from the leftover
portions of the four panels which were cut for desiccator test specimens and also from the fifth panel sampled for each product type.

Most of the products were tested relatively soon after
manufacture and again after they were aired out for 3 or more months
by placing them on stickers to allow air circulation between individual panels. The time allowed for airing is to be distinguished
from panel age or ageing, since formaldehyde levels tend to remain
constant for panels which are stacked tightly together; whereas,
levels decrease quite rapidly during airing out (10). Thus, airing
time is more critical than actual panel age when considering formaldehyde emissions. Since the panels for each of the products studied
were kept stacked together until they were conditioned for the
initial testing, all products were relatively "fresh", in one sense,
regardless of the time which had elapsed since their manufacture.

In the dynamic chamber tests, the large chamber (55.4 m^3) was
loaded at a rate of 0.43 m^2/m^3, and the ventilation rate was maintained at 0.5 air changes per hour. The test temperature was 25 $\pm$
1°C, and the relative humidity was held at 50 $\pm$ 5%. Air sampling
was accomplished with three sets of double impingers at one liter
per minute for 60 minutes, twice per day for two days. Formaldehyde
was analyzed using the acetylacetone procedure (10).

The results of the study are summarized in Table III which
provides 2-hour desiccator values and dynamic chamber values for
both fresh and aired out panels. For most of the product types,
both empty chamber and loaded chamber formaldehyde values are provided, the empty chamber values representing "background" measurements taken just before the chamber was loaded. These background
levels represent residual formaldehyde present in the chamber from
previous testing.

The data indicate that the loaded chamber values were below
0.1 ppm for all products and, also, that the background levels (in
the empty chamber) were on the same order of magnitude as the levels
observed with panels present. In fact, the data show that the background levels in some cases were as high as the levels measured when
the chamber was fully loaded, especially after the panels had been

Table III. Summary of Formaldehyde Test Data for Various Phenolic-Bonded Panel Products

Products(1)	2 Hour Desiccator (μg/ml)(2)	Large-Scale Dynamic Chamber (3)					
		Initial Test			Re-Test		
		Panel Age (Days)	W/Panels (ppm)(4)	Empty (ppm)(4)	Storage Time (Mo.)	W/Panels (ppm)(4)	Empty (ppm)(4)
SPP, 13 mm, 4-ply	0.08	32	0.04(5)	0.01	4	0.03	0.02
DFP, 14 mm, 5-ply	0.18	1	0.05	0.01	8	0.05	0.05
OSB No. 1, 12 mm	0.14	19	0.07	0.01	--	--	--
OSB No. 2, (Sample 1), 12 mm	0.02	21	0.07	0.07	3	0.04	0.02
OSB No. 2, (Sample 2), 12 mm	0.09	21	0.03	--	--	--	--
WB (Sample 1), 12 mm	0.17	55	0.08	0.03	8	0.01	0.01
WB (Sample 2), 12 mm	0.03	21	0.06	0.03	--	--	--
PB, 19 mm, hot-melt coating	0.15	16	0.08	--	9	0.03	0.01

(1) SPP = southern pine plywood; DFP = Douglas-fir plywood; OSB = oriented strand board; WB = waferboard; PB = particleboard.

(2) Micrograms of formaldehyde per ml distilled water. In this test, 25 ml distilled water in a petri dish is placed in a desiccator with eight 7 cm x 12.7 cm specimens for 2 hours, and the water is then analyzed for formaldehyde (16).

(3) Test conditions: 25 ± 1°C, 50 ± 5% RH, 0.5 air changes per hour, 0.43 m^2/m^3 loading (panel surface area/volume).

(4) Amount of formaldehyde, parts per million parts of air in the test chamber, volume basis.

(5) The sample was retested at 0 air changes per hour, and formaldehyde concentration was found to be 0.06 ppm.

aired. Because of the complex equilibria involved, it is not possible to simply correct the panel test data for background levels. Therefore, it is not feasible to use the results to derive an exact emission value for any of the products or to compare the various panel products. Instead, it is probably most prudent to simply conclude from this study that the upper limit on emissions from all types of phenolic panels, as determined in the large test chamber, is less than 0.1 ppm.

Although the high levels of "noise" due to background formaldehyde levels preclude meaningful statistical comparisons between emissions for various product types, certain other comparisons can be made. For example, a statistical t-test involving appropriately paired observations indicates a significant difference at the 1% confidence level between loaded and empty chamber values for the fresh panels. This difference indicates that the panels were probably contributing some formaldehyde to the test chamber, although it is not possible to determine how much, due to the complex equilibria involved. A similar analysis of the difference between loaded and empty chamber values for aired panels, however, shows a barely significant t-value at the 5 percent confidence level. Thus, the aired panels were probably contributing very little, if any, formaldehyde to the ambient atmosphere in the chamber. Evidently, the small amount of formaldehyde present initially in phenolic panels dissipates as the panels air out, so that loaded chamber levels approach background levels.

If background levels are ignored, a t-test involving paired observations representing those 5 sets of panels that were tested both before and after the panels were aired indicates that fresh panels emit more formaldehyde than aired panels (5% confidence level). Although such a statistical comparison is tenuous because of the confounding effects of the background levels, it is supported by the conclusions drawn above -- i.e., that fresh panels were apparently increasing the levels of formaldehyde in the chamber to a significant degree, while aired panels were contributing very little, if any, formaldehyde to the chamber.

The two-hour desiccator values shown in Table III are similar to those associated with the studies cited earlier in this report, and they are also indicative of extremely low formaldehyde emissions from the panels.

Other Unpublished Data. Table IV summarizes additional emission data which have been supplied to the American Plywood Association by various phenolic panel manufacturers. Data from both large-scale dynamic chamber tests and 2-hour desiccator tests are provided. This information agrees with that provided in the study described above and again demonstrates that formaldehyde emissions from phenolic panels are extremely low. In fact, for most of the products, the chamber background levels were as high as the levels during testing, suggesting that the products probably were not even contributing any formaldehyde to the chamber environment. These data again demonstrate that phenolic panels are such weak emitters that background formaldehyde levels can easily interfere with testing.

Table IV. Results of Large-Scale Dynamic Chamber Tests and Two-Hour Desiccator Tests on Various Types of Phenolic Panel Products(1)

Company Code	Product Type(2)	Panel Age at Test (Days)	Precondition Time(3) (Days)	Dynamic Chamber Test Parameters			Formaldehyde Levels		
				Loading Rate(4) (m²/m³)	Temp. (C)	Rel. Humid. (%)	Dynamic Chamber Empty (ppm)	Dynamic Chamber Loaded (ppm)	Two-Hr. Desiccator (μg/ml)(5)
A	SPP, 18mm, 4-ply	--	8	0.95	23 ± 0.5	48 ± 1	0.015	0.022	--
A	SPP, 16mm, 5-ply	--	8	0.95	23 ± 0.5	48 ± 1	0.010	0.020	--
A	SPP, 16mm, 5-ply	--	8	0.95	23 ± 0.5	48 ± 1	0.005(6)	0.011(6)	--
A	DFP, 13mm	--	8	0.95	23 ± 0.5	48 ± 1	0.013	0.017	--
B	PB, 16mm	--	2	0.49	24 ± 1	50 ± 1	--	0.04	0.51
C	SPP, 16mm, 5-ply	30	2-3	0.49	24 ± 1	50 ± 5	0.03-0.04(7)	0.03	0.09
C	COMPLY, 16mm	30	2-3	0.43	24 ± 1	50 ± 5	0.03-0.04(7)	0.04	0.13
C	WB, 16mm	22	2-3	0.52	24 ± 1	50 ± 5	0.03-0.04(7)	0.03	0.17
C	WB, 16mm	--	2-3	0.52	24 ± 1	50 ± 5	0.03-0.04(7)	0.05	0.18
D	PB, 19mm	--	2-3(8)	0.43	24 ± 1	45 - 55	0.01	0.04	0.22
D	PB, 19mm	--	7(9)	0.43	24 ± 1	42 - 48	0.01	0.05	0.17
D	PB, 19mm	--	7(10)	0.43	24 ± 1	42 - 50	0.01	0.04	0.20
E	SPP, 12mm	--	2	0.43	25 ± 1	42 - 56	--	0.04(12)	0.34(11)

(1) Ventilation rate was 0.5 air changes per hour for all tests. Chromotropic acid was used for HCHO analyses, unless noted otherwise. Different test chambers were used by each of the companies represented.

(2) SPP = southern pine plywood; DFP = Douglas-fir plywood; PB = particleboard; WB = waferboard.

(3) Specimens were preconditioned at the same temperature and relative humidity as is given for the test chamber.

(4) Loading is given in terms of square meters of panel surface per cubic meter of air volume in the chamber.

(5) All desiccator tests were performed in accordance with the procedures given by Test Method FTM 1-1983 (16) with edges unsealed unless noted otherwise. Values represent micrograms formaldehyde per ml distilled water. In this test 25 ml distilled water in a petri dish is placed in a desiccator with eight 7 cm x 12.7 cm specimens for 2 hours, and the water is then analyzed for formaldehyde.

Table IV. (continued)

(6) Pararosaniline was used for formaldehyde analysis, rather than chromotropic acid, using the same panel specimens as those used in the test whose results are reported directly above.

(7) Range typically encountered at this test facility.

(8) Background formaldehyde level in conditioning area was 0.08 ppm.

(9) Background formaldehyde level in conditioning area was 0.03 ppm.

(10) Background formaldehyde level in conditioning area was 0.05 ppm.

(11) Average of four tests involving samples from two separate panels: for each panel, samples for one test had sealed edges, while those for the other tests were unsealed. Range = 0.29 - 0.43.

(12) Average of four measurements made on four consecutive days. Range was 0.02 - 0.05.

Summary and Conclusions

All the available information indicates that formaldehyde levels
associated with wood panel products bonded with phenol formaldehyde
adhesives are extremely low. Data resulting from laboratory studies
involving large-scale dynamic test chambers consistently indicate
that levels are below 0.1 ppm under conditions simulating those
which might be found in tightly sealed homes containing freshly-
manufactured panels. In fact, test chamber levels are generally
about the same as the annual average formaldehyde concentrations
which have been reported for outdoor air in many cities (17). Moreover, the data, as well as theoretical considerations, indicate that
the amount of formaldehyde contributed to the environment by phenolic
panel products should rapidly approach zero as the small amount of
formaldehyde initially present in the panels is released.

Literature Cited

1.   Blomquist, R. F. "Formaldehyde Emissions Are No Problem With
      Wood Products Bonded With Phenolic Resins"; American Plywood
      Association: Tacoma, WA, 1981.
2.   Troughton, G. E. Wood Science 1969, 1, 172-6.
3.   "Formaldehyde; Determination of Significant Risk; Advance
      Notice of Proposed Rulemaking and Notice," Fed. Register,
      1984, 49, 21820-98.
4.   Nestler, F. H. Max. "The Formaldehyde Problem in Wood-Based
      Products -- An Annotated Bibliography"; U. S. Dept.
      Agriculture: Forest Products Laboratory, Report FPL-8, 1977.
5.   Berg, A.; Mellegard, B.; Hanetho, P.; Ormstad, E. B. Holz als
      Roh-und Werkstoff 1980, 38, 251-5.
6.   Couch, B. "Formaldehyde Study -- Testing of Plywood Panels";
      Weyerhaeuser Company: Tacoma, WA, Report 0452-102 IN, 1981.
7.   Myers, G. E. For. Prod. J. 1985, 35, 20-31.
8.   Myers, G. E.; Nagaoka, M. Wood Science 1981, 13, 140-50.
9.   "Large-Scale Test Method For Determining Formaldehyde Emissions
      From Wood Products -- Large Chamber Method, FTM 2-1983";
      National Particleboard Association: Gaithersburg, MD, 1983.
10.   Myers, G. E. For. Prod. J. 1983, 33, 27-37.
11.   Myers, G. E. For. Prod. J. 1984, 34, 59-68.
12.   "Formaldehyde Emission Comparison"; Champion International
      Corp.: Stamford, Conn., 1983.
13.   Meyer, B. "Formaldehyde Release From Phenolic Bonded Wood
      Panels"; American Plywood Association: Tacoma, WA, 1981.
14.   Sundin, B. Proc. 12th Wash. State Univ. International Sympos.
      on Particleboard, 1978, p.251.
15.   Roffael E. Proc. 12th Wash. State Univ. International Sympos.
      on Particleboard, 1978, p.233.
16.   "Small Scale Test Method For Determining Formaldehyde Emissions
      From Wood Products -- Two Hour Desiccator Test, FTM1-1983";
      National Particleboard Association: Gaithersburg, MD, 1983.
17.   "Indoor Pollutants," National Research Council, National
      Academy Press: Washington, D.C., 1983, p. 93.

18. Matthews, T. G.; Hawthorne, A. R.; Daffron, C. R.; Reed, T. J.;
     Corey, M. D. Proc. 17th Wash. State Univ. International
     Particleboard/Composite Materials Sympos., 1983, p. 179.
19. Matthews, T. G.; Hawthorne, A. R.; Daffron, C. R.; Reed, T. J.
     Proc. Air Pollution Control Assoc. Specialty Conf. on Meas. and
     Monitoring of Noncriteria Toxic Contaminants, 1983, p. 150.
20. "Formaldehyde Emission Survey of the Waferboard Association";
     National Particleboard Association: Gaithersburg, MD, 1984.
21. Roffael, E.; Mehlhorn, L. Holz als Roh-und Werkstoff 1980,
     38, 85-8.

RECEIVED January 14, 1986

4

Formaldehyde Release Rate Coefficients from Selected Consumer Products

J. A. Pickrell[1], L. C. Griffis[1,3], B. V. Mokler[1,4], C. H. Hobbs[1], G. M. Kanapilly[5], and A. Bathija[2,6]

[1]Lovelace Inhalation Toxicology Research Institute, Albuquerque, NM 87185
[2]U.S. Consumer Product Safety Commission, Washington, DC 20201

Formaldehyde (CH_2O) release was measured for seven
types of consumer products: pressed wood, urea
formaldehyde foam materials, clothes, insulation,
paper, fabric, and carpet. A modified Japanese
Industrial Standard (JIS) desiccator test was used
to measure release rate coefficients and to rank 53
products. Ten pressed wood products and five urea
formaldehyde foam products showed the highest CH_2O
releases (1–34 $mg \cdot m^{-2} \cdot day^{-1}$). The remainder,
representing all product types, had lower releases
ranging from 680 $\mu g \cdot m^{-2} \cdot day^{-1}$ to nondetectable
levels. In other studies, CH_2O release was measured
in a ventilated chamber for single samples of
particle board, plywood, insulation, and carpet.
When the combined CH_2O release was measured with both
particle board and one other product type (plywood,
insulation, or carpet) in the chamber, the values
obtained were less than the sum of that released
when each product was tested individually. This
finding suggested that CH_2O released from particle
board was reabsorbed by the second product (plywood,
insulation or carpet) being tested.

Many consumer products containing formaldehyde-based resins
release formaldehyde, leading to consumer annoyance and
health-related complaints (1–8). This release has led to various
symptoms, the most common of which are irritation of the eyes and
of the upper respiratory tract (2–5). Formaldehyde also produced
nasal carcinomas in mice and rats after exposure to 14.1 and 5.6

[3]Current address: Chevron Environmental Health Center, Richmond, CA 94804
[4]Current address: Small Particle Technology, Albuquerque, NM 87111
[5]Deceased
[6]Current address: U.S. Environmental Protection Agency, Washington, DC 20460

0097–6156/86/0316–0040$06.00/0
© 1986 American Chemical Society

ppm of formaldehyde, respectively, for long periods of time
(2-6). These findings have led to an intensified interest in
formaldehyde release from various consumer products into the
indoor environment. Consumer products, specifically construction
materials, are a major source of formaldehyde in the indoor
environment (7). Little information is available concerning
formaldehyde release from various consumer products.

In these studies, formaldehyde release rate coefficients
were measured for different consumer products using two methods.
In one series of studies, a small static chamber with no
ventilation, which was a modification of the Japanese Industrial
Standard (JIS) desiccator procedure, was used to compare
formaldehyde release from a number of products (1, 7-14). In a
second series of studies, a chamber with ventilation rates similar
to those in houses was used to more closely mimic actual product
use. With this method, combined formaldehyde release from two
products placed in the same chamber was compared to their separate
releases.

Materials and Methods

Desiccator Measurements. Fifty-three different brands or lots of
consumer products of seven different general types were analyzed
in this study (Table I). All but two of the wood products, and
the samples of urea formaldehyde foam, were purchased from
commercial sources by the Consumer Product Safety Commission. The
two wood products were purchased locally and are so identified.
Samples of urea formaldehyde foam (UFF) were provided by Drs.
Keith Long and Clyde Frank of the University of Iowa (Iowa City,
IA). At this time Drs. Long and Frank also provided samples of
drywall which had been placed next to urea formaldehyde foam for
more than 1 week in a configuration like that of a building. This
drywall was analyzed to determine the degree to which it had
absorbed formaldehyde from the UFF and subsequently released
formaldehyde under our test conditions. The time of manufacture
of the products relative to acquisition was not known. After
acquisition, samples were encased in plastic wrap until
conditioning to minimize release of formaldehyde prior to testing
(3 to 9 mo. after acquisition).

Table I.   Samples Analyzed by the Modified JIS Desiccator
Procedure

General Types of Samples	No. of Different Samples Analyzed
Pressed Wood Products	12
Urea Formaldehyde Foam Products	7
New Unwashed Clothes	4
Insulation Products	6
Paper Products	3
Fabric	14
Carpet	7

Each of these products was conditioned at room temperature, and ~ 100% relative humidity (RH) (31 to 67 days). Formaldehyde release was measured as described (1, 8, 15). A modified JIS desiccator procedure was used, and formaldehyde was quantitated using a pararosaniline procedure (15, 16). Formaldehyde release rate coefficients were calculated (15). An average coefficient of variation of 16% was obtained for this measurement (15). Samples displaced less than 12% of the chamber air (15).

Dynamic (Ventilated) Chamber Measurements. One sample each of particle board, plywood, insulation material, and carpet was tested. The U. S. Consumer Product Safety Commission, Bethesda, MD purchased these samples. Formaldehyde release was measured in a dynamic (ventilated) chamber system with one air change per hour as described (17). Air temperature and humidity were controlled. Formaldehyde was trapped using a midget impinger train (17). Samples displaced less than 12% of the chamber air (17). Aqueous formaldehyde and total extracted formaldehyde were measured as described (1, 8, 15-18).

After testing each of the four individual products, three pairs of products were tested. Formaldehyde release when multiple products were in the same chamber was measured as above. The three pairs tested were particle board/plywood, particle board/insulation, and particle board/carpet.

Results

As measured by the modified JIS desiccator procedure, pressed wood products had the highest release rate coefficients expressed as a function of surface area (Table IIA) of the various sample types tested. Release rate coefficients from urea formaldehyde foam products were comparable to those of pressed wood products (Table IIB). Products labelled substrate (sub 1, sub 2, and sub 6) were experimental foams. The drywall that had been placed next to the foams (Number 1, 2, or 3) for more than 1 week in a configuration similar to that in a building released a moderate amount of formaldehyde.

Unwashed new clothing samples (Table IIC), fiberglass insulation products with formaldehyde resins (Table IID), paper products (Table IIE), fabrics (cotton, nylon, olefin, and blended) (Table IIF), and carpets (Table IIG), had substantially (≈ 3 to > 100 fold) lower formaldehyde release rate coefficients, as measured by this method, than did pressed wood products or urea formaldehyde foams (1, 15).

If one ranks the various consumer products in this survey based on their release coefficients per unit of surface area, more than 45% of the samples (24 samples) had very low offgassing rate coefficient (< 100 μg of formaldehyde released (m^2 of product surface area)$^{-1}$ day^{-1}). Six of seven categories of products tested had individual samples with these low offgassing rates. Less than one-third of the samples (15 samples) had offgassing rate coefficients greater than 1000 μg m^{-2} day^{-1} (Table II) (1, 15).

No consistent differences were observed between release rates from products measured in ventilated chambers and

Table II. Release of Formaldehyde from Consumer Products

	$\mu g\ g^{-1}\ day^{-1}$	$\mu g\ m^{-2}\ day^{-1}$
(A) Pressed Wood Products		
Particle board		
A	4.1-5.3[a]	13000-17000[b]
B	6.7-8.1	23000-26000
C	4.9-7.1	20000-28000
D	0.4-0.4	1800-2200
Plywood		
A (interior)	7.5-9.2	13000-15000
B (exterior)	0.03-0.03	54-56
C (exterior)	ND (0.01)[c]	ND
Paneling		
A	19-21	32000-36000
B	4.6-4.7	7100-7500
C	6.9-7.3	6400-6900
D	3.9-4.3	5200-5600
E	0.84-0.86	1480-1540
(B) Urea Formaldehyde Foam Insulation Products		
Urea formaldehyde foam		
1	59-67	22000-28000
2	54-54	12000-14000
3	53-67	18700-18800
sub 1	25-31	5400-7500
sub 2	88-91	21000-22000
sub 6	ND (.1075)[c]	ND
Drywall		
1	0.10-0.16	400-600
(C) New Clothes		
Men's shirts (polyester/cotton)	2.5-2.9	380-550
Ladies' dresses	3.4-4.9	380-750
Girls' dresses (polyester/cotton)	0.9-1.1	120-140
Children's clothes (polyester/cotton) 15-55		0.2-0.3
(D) Insulation Products		
Fiberglass ceiling panel 0.75-in.	1.3-1.7	390-540
Rigid round airduct	0.66-0.72	390-430
Rigid round fiberglass duct	0.06-0.06	150-150
Fiberglass	1.0-2.3	260-620
Fiberglass 3.5-in.	0.3-0.7	52-130
Blackface insulation sheathing	0.03-0.04	340-420

Continued on next page

Table II (Continued)

(E) Paper Plates and Cups		
A	0.12-0.36	400-1000
B	0.03-0.14	75-450
C	0.10-0.15	330-335
(F) Fabrics		
Drapery fabric		
A (100% cotton)	2.8-3.0	330-350
B (100% cotton)	0.8-0.9	90-120
C (blend, 77% rayon-23% cotton)	0.3-0.3	50-50
D (blend, 77% rayon-23% cotton)	ND (0.01)[c]	ND
Upholstery fabric		
A (100% nylon)	0.03-0.05	9-11
B (100% nylon)	0.02-0.02	6-7
C (100% olefin)	0-0.02	0-5
D (100% olefin)	ND (0.014)	ND
E (100% cotton)	ND (0.014)	ND
F (100% cotton)	ND (0.015)	ND
Latex-backed fabric		
A	0.5-0.6	90-100
B	ND (0.015)	ND
Blend fabric		
A	0.3-0.4	20-30
B	0.2-0.3	20-30
(G) Carpets		
A (foam-backed)	0.05-0.06	60-65
B (foam-backed)	0.006-0.01	8-13
C (foam-backed)	0-0.002	0-2
D	0.0005-0.0009	0-4
E	0.0007-0.0009	0-1
F	0-0.0009	0-1
G	ND (0.043)[c]	ND

[a]Range of two or more measured values expressed as µg of formaldehyde (g of product)$^{-1}$ day^{-1}.
[b]Range of two or more measured values expressed as µg of formaldehyde (m^2 of area product)$^{-1}$ day^{-1}.
[c]ND = below limit of detection. Parentheses contain limit of detection.

nonventilated desiccators between loadings of 1.4 and 21 m^2/m^3 (17). Release rate coefficients measured in ventilated chambers at 9-11 m^2/m^3 differed by 13% from release rate coefficients analyzed under modified JIS desiccator conditions (nonventilated) for the same products when extrapolated to a loading of 21 m^2/m^3 (Table III) (17). Release rate coefficients for particle board or plywood measured in a ventilated chamber at a loading of 1.4-1.6 m^2/m^3 were 4-33% different from those measured in a desiccator at similar loadings (1.4-1.8 m^2/m^3). A similar comparison indicated that release rate coefficients for particle board plus plywood measured in the ventilated chamber were 14% higher than those measured in the desiccator at loadings of 3.0-3.4 m^2/m^3 (Table III).

In dynamic (ventilated) chambers, release rate coefficients were increased by a factor of 4.4 for particle board and 2.2 for plywood at loadings of 1.4-1.6 m^2/m^3 over values at loadings of 9-11 m^2/m^3 (Table IV). Increased pressure of formaldehyde in the chamber was associated with reduced release of formaldehyde from wood products, as indicated by comparing equilibrium concentrations of formaldehyde (17).

Formaldehyde release rates were measured using multiple consumer products in a dynamic chamber. Particle board and plywood had high formaldehyde specific release rate coefficients. Combined plywood and particle board had a release rate 68% of the sum of the two products and 91% of the particle board release (Table V). When particle board was combined with insulation, the combined release rate was ~ 71% of the sum of the separate release rates and 73% of the particle board release. Particle board and carpet combinations gave similar results.

A good correlation was noted between release rate coefficients at loadings of 1.4-2.8 m^2 of product surface area/m^3 of chamber volume and formaldehyde extractable into toluene (Table V; r^2 = 0.999; p = < 0.001). Total extractable formaldehyde was quite low in both carpet and fiberglass insulation (0.5-1.6 mg/100 g of material) relative to that in plywood or particle board (22-55 mg/100 g of material) (Table V) (17).

Discussion

Pressed wood products and urea formaldehyde foam products had much higher release rates than those from most of the other products tested. Similar release rates have been observed by others (19). More than half of the products tested had very low release rate coefficients, and this included individual samples from six of seven of the types of products. Products equilibrated at 100% RH prior to the measurement were used to measure formaldehyde release. This equilibration may have removed a variable amount of formaldehyde (8, 14-17).

The relative ranking for each type of product on the basis of rate of release of formaldehyde per unit surface area was pressed wood products ~ urea formaldehyde foam >> clothes ~ insulation products ~ paper products > fabric > carpet. Considering the surface area of each type of product likely to be present in houses and the relative release rate coefficients,

Table III. Comparison of Formaldehyde Release Rate Coefficients in Ventilated Chambers and Nonventilated Desiccators

| | Loading (m^2/m^{3a}) | | Release rate Coefficient ($\mu g\ m^{-2}\ day^{-1}$) | | |
	Ventilated Chamber	Nonventilated Desiccator	Ventilated Chamber	Nonventilated Desiccator	Difference %
Particle board	21	21	21000[b]	24000	13
Plywood	21	21	16000[b]	14000	13
Particle board	1.4	1.8	168000	121000	33
Plywood	1.6	1.5	68000	71000	4
Particle board + plywood	3.0	3.4	71000	65000	14

[a]Chamber loading in m^2 of product surface area/m^3 of chamber volume.
[b]Corrected to a loading of 21 m^2/m^3 by using data in Table III.

Table IV. Loading Effect of Plywood and Particle Board at ~ 25°C and ~ 90°C RH in Ventilated Chamber[a]

	Extractable Formaldehyde (mg/100 g)	Loading[b] (m^2/m^3)	Formaldehyde Release Rate Coefficient[c,d] ($\mu g\ m^{-2}\ day^{-1}$)	Calculated Loading Effect[e]
Particle board	55	11	38000	4.4
		1.4	168000	
Plywood	22	8.6	31000	2.2
		1.6	68000	

[a]One air change per hour was the flow rate.
[b]m^2 of product surface area/m^3 of chamber volume.
[c]μg of formaldehyde released (m^2 of surface area of product)$^{-1}$ day^{-1}.
[d]Offgassing strengths of formaldehyde extrapolated to a loading of 21 m^2/m^3 were 21000 (particle board) and 16000 $\mu g\ m^{-2}$ day^{-1} (plywood).
[e]This number represents the ratio of the release rate coefficient at low loading compared to high loading.

Table V. Release Rate Coefficients from Product Combinations in Dynamic Chambers

Sample	Loading[a] (m²/m³)	Formaldehyde Release Rate Coefficient[b] ($\mu g\ m^{-2}\ day^{-1}$)	Total Formaldehyde Release Rate[c] ($\mu g\ day^{-1}\ m^{-3}$)	Total Extractable Formaldehyde (mg/100 g)
Particle board	1.4	168000	235000	55
Plywood	1.6	68000	109000	22
Particle board and plywood	3.0	71000	213000	ND[d]
Insulation	1.7	3000	5000	1.6
Particle board and insulation	3.1	55000	171000	ND
Carpet	2.8	1500	4000	0.5
Particle board and carpet	4.2	31000	129000	ND

[a] m² of surface area/m³ of chamber volume.
[b] μg of formaldehyde released (m² surface area product)⁻¹.
[c] This number is the multiple of the first two columns: (Formaldehyde release rate coefficient) (loading).
[d] ND = not determined

pressed wood products and urea formaldehyde foam appear to have
the greatest potential for formaldehyde release in a house.

Release rate coefficients determined in this report for a
variety of products are only one way of assessing the relative
potential for release of formaldehyde from these products. The
release rate coefficient based on surface area is a more realistic
measure of potential release than is one based on weight. In this
report, samples were measured at loadings of 21 m^2 surface
area/m^3 chamber volume. Values of greater than fivefold higher
for release of formaldehyde were measured for particle board and
plywood at lower loadings of 1.5-1.8 m^2 of surface area/m^3 of
chamber volume (1, 5, 17). The degree to which the ranking in
this report would change under loading conditions more like the
conditions typically present in houses and mobile homes should be
investigated.

Formaldehyde release rate coefficients measured in
desiccators were similar to those determined in the dynamic
chamber at similar loadings. Initial formaldehyde release rate
coefficients for one sample each of particle board and plywood
measured at 11.4 and 8.6 m^2/m^3 in these chambers at one volume
change per hour were ~ 2-fold higher than those measured in
desiccators at higher loadings (8, 15, 17). However, when the
release rate coefficients were adjusted for differences in
loading, the calculated release rate coefficients were similar to
those measured in desiccators (8, 15, 17).

Particle board and plywood released sufficient formaldehyde
in the dynamic chambers to attain air concentrations that
approached calculated equilibrium air concentration values. At
9-11 m^2/m^3 loadings, concentrations of formaldehyde were
> 50% of calculated equilibrium concentrations, probably because
airflow was low relative to the mass of the product. The high
chamber concentration of formaldehyde may have limited
formaldehyde release in the dynamic chamber.

Reduced sample loadings in the dynamic chamber led to
decreased formaldehyde concentrations in the chamber as noted or
predicted previously by others (17, 20-22). This resulted in
increased release rate coefficients (μg m^{-2} day^{-1}). Samples
analyzed at 1.4 and 1.6 m^2 of product surface area/m^3 of
chamber volume chamber loadings had formaldehyde chamber
concentrations of 28-32% of the calculated equilibrium air
concentrations of formaldehyde (17), suggesting better relative
ventilation than that at higher chamber loadings.

When particle board was paired with plywood, insulation, or
carpet and tested in a dynamic chamber, the formaldehyde released
was ~ 60% of the sum of that released when each product was
tested alone. Similar results have been observed by others (19).
Approximately half of this reduction is related to the increase in
chamber loading noted in Table IV (14-17). In fact, the release
of formaldehyde when these products were combined with particle
board was less than that released by particle board alone. These
results suggest that formaldehyde from the high-emitting particle
board moved into the lower emitting product. If this is the case,
it is highly likely that the water present in the second product
actually absorbed some formaldehyde given off by the particle
board since formaldehyde tends to move into the water phase of the

product (23). To confirm this, corroborating measurements would
be necessary. The water in the low-emitting product may act as a
sink to absorb formaldehyde from the high-emitting product and
reduce formaldehyde concentrations in a room by ~ 30 -50%. Wood
contains approximately the same amount of water as pressed wood
products and might behave in the same way. This factor would
become important in houses where surface areas of pressed wood
products were small compared to that of other wood. Most houses
contain large surface areas of carpet or insulation relative to
that of pressed wood products. The former products may account
for substantial reductions in total formaldehyde concentrations
when used with pressed wood products (17).

Summary

Most products tested released only small amounts of formaldehyde.
Only some pressed wood and urea formaldehyde foam insulation
products released higher amounts of formaldehyde. Products tested
in both ventilated chambers and unventilated desiccators released
similar amounts of formaldehyde. Formaldehyde released by
particle board was reabsorbed by the second product tested in a
dynamic chamber. In a house this reabsorption might lower the
room level of formaldehyde.

Acknowledgments

Research performed for the Consumer Product Safety Commission
under Interagency Agreement CPSC-IAG-801463 under U. S. Department
of Energy Contract DE-ACO4-76EVO1013. The views expressed in this
paper are those of the authors and do not necessarily represent
those of The Consumer Product Safety Commission or the U. S.
Department of Energy. The technical assistance of Tina Shifani
and other members of the ITRI staff, the illustrative assistance
of Emerson E. Goff, and the scientific and editorial review of A.
Dahl, J. Bond, S. Rothenburg, R. Henderson, and other members of
the ITRI staff, as well as A. Ulsamer and W. Porter of the
Consumer Product Safety Commission, are gratefully acknowledged.

Literature Cited

1.   Meyer, B. "Urea Formaldehyde Resins"; Addison-Wesley:
      Reading, MA, 1979; p. 423.
2.   Blackwell, M.; Kang, H.; Thomas, A.; Infante, P. Am. Ind.
      Hyg. Assoc. J. 1981, 42(7), A34-A46.
3.   Albert, R.; Sellakumar, A.; Laskin, S.; Kuschner, M.;
      Nelson, N.; Snyder, C. A. J. Natl. Cancer Inst. 1982, 68,
      597-604.
4.   Committee on Toxicology "Formaldehyde--An Assessment of Its
      Health Effects," prepared for Consumer Product Safety
      Commission by National Academy Sciences, 1980.
5.   Moschondreas, D. J.; Reactor, H. E. "Technical Report LBL
      12590, EEB-Vast 81-12," National Technical Information
      Service, 1981.

6. Swenberg, J. A.; Kerns, W. D.; Mitchell, R. I.; Gralla, E. J.; Pavkov, R. L. Cancer Res. 1980, 40, 3398-3401.
7. Committee on Indoor Pollutants "Indoor Pollutants"; National Research Council, National Academy Press: Washington, DC, 1981.
8. Pickrell, J. A.; Griffis, L. C.; Hobbs, C. H. "Final Report to the Consumer Product Safety Commission," Lovelace Inhalation Toxicology Research Institute, Albuquerque, NM, LMF-93, National Technical Information Service, 1982.
9. Meyer, B.; Koshlap, K. "LBL Report 12570 (DRAFT)," Molecular and Materials Research Division, Lawrence Berkeley Laboratory, University of California, Berkeley, CA 1981.
10. Myers, G. E.; Nagoka, M. Symposium on Wood Adhesives--Research, Applications and Needs, Forest Product Laboratories, Madison, WI, Sept 1980.
11. Myers, G. E.; Nagoka, M. Wood Sci. 1981, 13(3), 140-150.
12. Sundin, B. Presented at The International Particle Board Series Symposium No. 16, Washington State University, Pullman, WA, March 30-April 1, 1982.
13. Fujii, S.; Suzuki, T.; Koyagashiro, S. Kenzai Shiken Joho Transl., 1973, 9(3), 10-14.
14. Griffis, L.; Pickrell, J. A. Environ. Int. 1983, 9, 3-7.
15. Pickrell, J. A.; Mokler, B. V.; Griffis, L. C.; Hobbs, C. H.; Bathija, A. Environ. Sci. Technol. 1983, 17, 753-757.
16. Miksch, R. R.; Anthon, D. W.; Fanning, L. Z.; Hollowell, C. D.; Revzan, K.; Glanville, J. Anal. Chem. 1981, 53, 2118-2123.
17. Pickrell, J. A.; Griffis, L. C.; Mokler, B. V.; Kanapilly, G. M.; Hobbs, C. H. Environ. Sci. and Technol. 18, 682-686.
18. Tiffany, T. O. CRC Crit. Rev. Clin. Lab. Sci. 1974, 5, 129-191.
19. Singh, J.; Walcott, R.; St. Pierre, S.; Ferrell, T.; Garrison, S.; Gramp, G.; Groah, W. "Evaluation of the Relationship between Formaldehyde Emissions from Particle Board Mobile Home Decking and Hardwood Plywood Wall Panelling in Experimental Homes;" prepared for U.S. Department of Housing and Urban Development, Office of Policy Development and Research, Division of Energy, Building Technology and Standards, 1982.
20. Meyer, B. "Urea Formaldehyde Resins"; Addison-Wesley: Reading, MA, 1979; p. 423.
21. Myers, G. E.; Nagoka, M. Presented at the Symposium on Wood Adhesives--Research, Applications and Needs, Forest Product Laboratories, Madison, WI, 1980.
22. Esmen, N. A. Environ. Sci. Technol. 1978, 12, 337-339.
23. Johns, W. E.; Jahan-Latiban, A. Wood Fiber 1980, 12, 144-152.

RECEIVED January 14, 1986

5

Cellulose Reaction with Formaldehyde and Its Amide Derivatives

B. A. Kottes Andrews, Robert M. Reinhardt, J. G. Frick, Jr., and Noelie R. Bertoniere

U.S. Department of Agriculture, Southern Regional Research Center, New Orleans, LA 70179

Research establishing the reaction between cellulose and formaldehyde or formaldehyde adducts is reviewed. The reactions involve etherification of the accessible cellulose. The etherification has resulted in commercial modifications that are important to cellulosic textiles. Gross effects of the etherifications that crosslink cellulose in textiles are increased resiliency, manifested in wrinkle resistance, smooth-drying properties and greater shape-holding properties; and conversely, reduced extensibility, strength and moisture regain. Both chemical and physical evidence of the cellulose etherification are reviewed. Estimation of the degree of crosslinking for several agents including formaldehyde and urea-formaldehyde is presented as chemical evidence of cellulose reaction. Physical evidence of crosslinking can be seen in the response of the crosslinked fibers to cupriethylenediamine and to a methacrylate layer-expansion treatment that separates lamellae and reveals gross representations of the crosslinking effect.

Cellulose is the major component of cotton, wood, and many of the bast fibers such as linen, flax, ramie and jute and also the component that undergoes the most useful reactions. Although the microstructural units of the cellulose, impurities, and hence the accessibility to reagents, differ among these natural fibers, the chemical nature and reactivity are the same. By analogy, mechanisms established for cotton cellulose modifications should be valid for other celluloses. While there apparently is still controversy among wood chemists over whether crosslinking occurs in wood cellulose, the chemistry of crosslinkage of cellulose and other glucoses is well established by the research summarized in this chapter.

Because of consumer demand in the second half of this century for easy care textiles, interest in the reactivity of cellulose from the ever popular cotton and viscose rayon preceded interest in the other products. In fact, it is the alcohol functionality of cotton and viscose cellulose that is responsible for improvements in the aesthetic and functional properties of their fibers and fabrics.

Figure 1 shows the repeating glucose units of cellulose with the carbons labeled, including those with the reactive 2, 3, and 6 hydroxyls. The most important reactions of cotton cellulose commercially are esterification and etherification, with the products of etherification ranking first. It is generally agreed today among textile scientists that durable press cellulosic textiles owe their smooth-drying and resilient properties to the reactivity of formaldehyde and its amide derivatives with cellulose to produce crosslinks between adjacent cellulose chains (Figure 2). However, the theory that crosslinking was responsible for increased resiliency developed only after the treatments were in wide use.

Early Developments

The earliest reference to cellulose crosslinking was the work of Meunier and Guyot (2). Crosslinking to form methylene bridges was suggested as the mechanism for treatment of viscose rayon by an acid formaldehyde process. Although this "cross-bonding" theory was proposed by other workers in the following years, the hypothesis was not supported by experimental evidence.

Later treatments by other research workers used melamine-formaldehyde and urea-formaldehyde which gave less strength loss than did the treatments with formaldehyde itself (3,4). Because these agents form polymers and did cause less strength loss, they were considered polymer-formers or resin-formers rather than crosslinking agents, hence the term "textile resins". Cameron and Morton proposed that urea-formaldehyde, or methylolureas, did crosslink, but still considered that polymer formation was the most important part of the reaction (3). They estimated that, in a 15% materials add-on, that 1% was involved in crosslinking and 14% in polymerization.

Steele and Giddings showed that the composition of products from dimethylolurea on cotton indicated that crosslinking was the primary reaction for "crease-resist" properties; little polymer was formed although crosslinks contained more than one urea residue (5). Commercial products, however, were mixtures of monomethylol- and dimethylolurea, and were more likely to form polymers. As the crosslinking theory developed, crosslinking was established as the essential reaction for obtaining resiliency, while polymer formation was seen to affect other properties only, sometimes adversely.

Crosslink Theory Development

Although the first use of urea-formaldehyde in production of anti-crease textiles was patented by Tootal, Broadhurst, Lee Co., Ltd. in 1928 (6), crosslinking of cellulose with methylolamides was first proposed by Cameron and Morton in 1944 (3). They argued that cellulose crosslinking occurred with methylolureas on rayon, but considered polymerization also important for the desired anti-crease effects. Gagliardi and Nuessle, by analogy with physico-chemical evidence from other high polymers, suggested that the changes in chemical, physical and mechanical properties of cellulose effected by treatment with "wrinkle proofing" agents could be logically explained by crosslinking (7).

In landmark research, Cooke et al. presented the first chemical evidence for crosslinking (8). They showed that melamine formaldehyde treatments of cellophane films produced changes in the region of the

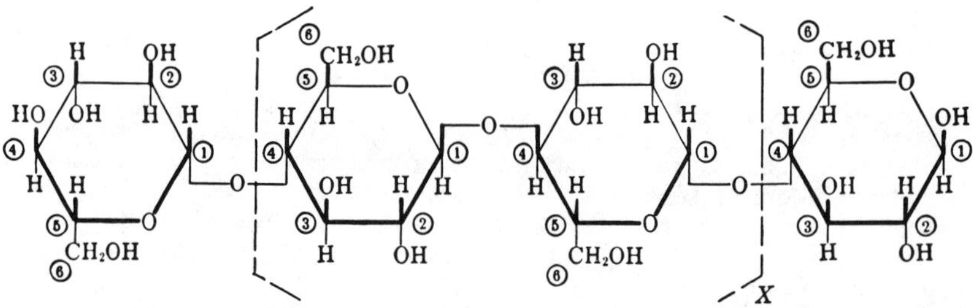

Fig. 1. Anhydroglucose units in the polymeric chain of cellulose (1).

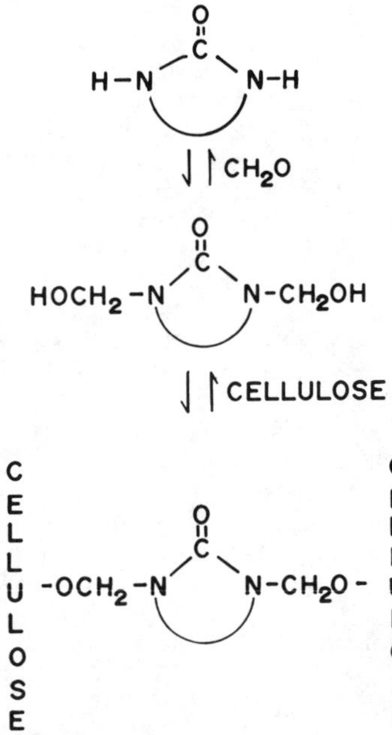

Fig. 2. Synthesis of a methylol agent and its reaction with cellulose (1).

infrared spectra of the films associated with the C-O bond stretching. These changes suggested formation of a cellulose-amidomethyl ether. Much later, in 1974, Madan used polarized infrared to show, with dimethylolethyleneurea and dimethyloldihydroxyethyleneurea, that reaction was intermolecular, not intramolecular, on cellulose (9).

Intermolecular crosslinking of cellulose by formaldehyde has also been established by chemical means. Rao, Roberts and Rowland isolated formaldehyde crosslinked constituents from ball-milled cotton cellulose modified with formaldehyde in a swollen state and subsequently hydrolyzed. Cellotriose oligomers joined through formal linkages and cellobiose pairs joined through formal linkages were identified from the hydrolysates of the disordered celluloses by paper chromatography (10).

By use of analyses for total nitrogen and formaldehyde contents of fabrics treated with formaldehyde and methylolamide cellulose reactants, the size of the crosslinks could be measured. Steele and Giddings found that the length of a crosslink from dimethylolurea contained 2.0 urea residues (5). Frick, Kottes and Reid confirmed this finding and extended the information to estimate ethyleneurea crosslinks at 1.3-1.4 ethyleneurea residues, and formaldehyde crosslinks to be monomeric (11). In addition, the crosslinks per anhydroglucose unit (agu) were calculated over a range of addons for these three reactants (Table I). Earlier work by these researchers had established that, in dimethylolethyleneurea treatments of cotton, crosslinking was the primary reaction; little, if any homopolymer formed (12).

Increases in resiliency and the corresponding losses in extensibility and strength have been related to the extent of crosslinking. It was found that, for dimethylolurea (DMU), dimethylolethyleneurea (DMEU), and formaldehyde (HCHO), maximum resiliency, as measured by wrinkle recovery angles, is attained at a substitution of 0.04-0.05 crosslinks per agu (Figure 3). This relationship between maximum resiliency and crosslink concentration was confirmed by Gardon (13). Values for the other physical properties also tend toward a maximum deviation from untreated fabric at this same substitution. Two factors were found to contribute to strength loss in crosslinked cotton fabric: reduction of extensibility, or stress distribution from crosslinking, and acid degradation of the cellulose by acidic catalysts. The former cause is common to all crosslinked fabrics, but the latter has a noticable effect with formaldehyde-crosslinked fabrics. High strength losses associated with formaldehyde crosslinking occur because it requires stronger acidic catalysis than does amidomethylol crosslinking (11).

Both the reactivity of the crosslinking agents to etherification of cellulose and resistance of these cellulose crosslinks to hydrolysis were found to depend on the electron density around the amidomethyl ether group, and thus, suggested a carbocation mechanism for reaction under acidic conditions. Attack on the ether oxygen by a positive ion facilitates cleavage at the C-O bond to give cellulose as an initial product of hydrolysis (14,15). In research to elucidate the chemical structure of crosslinked cottons by a sequential analytical scheme, Willard, et al., presented chemical evidence for

Table I. Crosslink Substitution on Cotton Fabric Finished for Wrinkle
Resistance (11).

Finishing agent	N %	HCHO %	Crosslinks per agu	Molar ratio HCHO residues per N/2
Dimethylol urea	0.17	0.19	0.0004	1.05
	0.54	0.85	0.015	1.45
	1.26	2.03	0.039	1.50
	2.65	4.33	0.087	1.53
Dimethylol ethyleneurea	0.11	0.16	0.003	1.35
	0.29	0.48	0.009	1.55
	0.57	1.10	0.027	1.80
	1.41	2.68	0.066	1.77
Formaldehyde	--	0.10	0.005	--
	--	0.26	0.014	--
	--	1.00	0.054	--
	--	1.73	0.095	--

involvement of some of each of the 2, 3, and 6 cellulose hydroxyls
(Figure 1) in covalent crosslinking (16). Also, the relative
reactivities of these hydroxyls of cellulose were claimed by Peterson
(17) and Vail (18) to influence the kinetics and thermodynamics of
cellulose etherification.

Some amidomethylol agents can also crosslink cellulose under
alkaline conditions. For such cases a different mechanism of reaction
and hydrolysis was proposed that favored initial cleavage at the C-N
bond to give a cellulose hemiacetal as an initial product of
hydrolysis (15). Although in most cases improvements in cotton
fiber/fabric resiliency by chemical treatment are produced by
crosslinkage of adjacent cellulose chains and not by polymerization,
there are some exceptions. Notable are the improvements in resiliency
imparted to cotton fabric by long chain fatty esters (19,20), by
deposition of crosslinkable polysiloxanes (21) and other elastic
polymers (22,23,24). It should be considered that in all of these
exceptions, the addon is much higher than that observed with finishes
from crosslinking agents. For example, McKelvey and his co-workers
report a DS of approximately 0.1 for four finishes from monofunctional
long chain acid chlorides. It should be noted that a DS of 0.1
required a high weight addon because of the high molecular weight of
the substituent. Electron photomicrographs showed that a smooth
polymer film had covered the fiber surface as a result of the
treatments (19). Bullock and Welch suggest that, with polysiloxanes,
an elastic covering forms over the individual fibers, and augments the
cotton fibers' inherent recovery forces. The term, "fabric coating"
is used (21). Steele and his co-workers offered a theory of inter-
yarn "spot welding" to explain contributions of these elastomers to
resiliency improvements (25), but this was shown not to occur (26).
Improvements in resiliency are more likely caused by the high energies
of extension and recovery in the polymer film itself (21,22).

Crosslinking Agent Development

Cellulose reactants have progressed throughout the years from the
early urea-formaldehyde, melamine-formaldehyde and phenol-formaldehyde
agents for wash-and-wear finishes to the modern methylolated cyclic
ureas for durable press as the durability requirements evolved.
Crease-proof finishes from methylolated ureas and melamines did not
withstand the common conditions of home and commercial launderings
(15). This instability precluded the finishes' use for shirting and
other fabrics routinely sent to commercial laundries in the custom of
the day. The discovery of methylolated imidazolidinone-2, or cyclic
ethyleneurea, provided an improved wash-and-wear garment with
aesthetic properties that survived commercial laundering (17).
    Another impediment to consumer acceptance of fabrics finished for
crease resistance was the lack of durability to chlorine bleaching.
While methylolated ethyleneurea finishes had good resistance to damage
from retained chlorine if applied properly, treatment factors, such as
degree of methylolation, choice of catalyst and degree of cure were
critical to a chlorine resistant finish (28,29). The search for a
replacement agent led to the use of dimethyloltriazones for crease
resistance in instances where chlorine resistance was necessary (30).
    Further refinements in agents for higher level crease resistant,
smooth drying cellulosic fabrics led to the development of
dimethyloldihydroxyethyleneurea (DMDHEU), the agent used to finish 80%
of the durable press fabrics today. Finishes from this agent combine
high performance with acid stability and chlorine resistance. In
addition, the use of DMDHEU allowed reduction in the amount of free
formaldehyde released by the agent and treated fabric. Formaldehyde
release levels in fabrics have been brought down from the 5000 µg
based on 1 g fabric routinely measured in the AATCC Test Method 112
(Sealed Jar) (31) with the first wash-and-wear fabrics to less than
500 µg based on 1 g fabric with the second and third generation DMDHEU
and methylolated carbamate agents in use today (32,33,34).
    Figure 4 lists the types of methylolated amides typically used as
cellulose reactants. However, formaldehyde release and the regulatory
response to potential consumer hazards from it (35) have led to a
search for formaldehyde free cellulose reactants. Whereas some are
departures from the typical amidomethylol chemistry successful for
cellulose crosslinking (36,37), the most widely used contain a
reactive hydroxyl alpha to an amido group as in the methylolated
agents (38,39,40). At best, formaldehyde free agents have limited
commercial use in the United States, mainly in baby clothes. Some
non-formaldehyde reagents such as 2-substituted amines, however, have
been quite useful in establishing the nature and position of
crosslinks between cellulose groups, both by chemical analysis of
modified cotton cellulose (41) and by synthesis of crosslinked
glucoses (42).

Crosslinking Response

Although not a measure of cellulose crosslinking, since monofunctional
agents are incapable of crosslinking, the response to hydrolysis
conditions of cotton fabric treated with N-methyl, N'-
hydroxymethylethyleneurea offers evidence of cellulose reaction. This
response can be seen in Table II. Formaldehyde is released from the

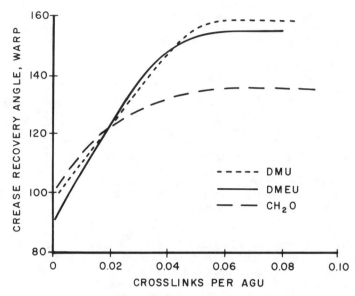

Fig. 3. Crease recovery angle of crosslinked fabrics (11).

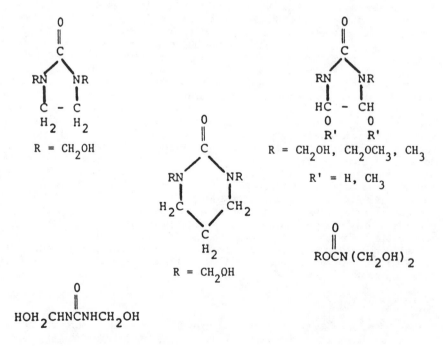

Fig. 4. Amido compounds used in production of commercial finishing agents.

Table II. Response of Cotton Printcloth Treated with N-methyl,N'-
hydroxymethylethyleneurea to pH Variation in the Japanese Law 112-1973
Test (43)

Formaldehyde release* (µg/g) after incubation at:		
pH 2	pH 7	pH 10
4198 (53)	3196 (1071)	332 (2162)

*Numbers in parentheses are the values obtained after the fabric
residues from the incubations at the indicated pH were given a second,
standard (pH 7), incubation in the Japanese 112-1973 test.

amidomethylether side chain on hydrolysis. The formaldehyde could
have come only from the hydrolyzed reaction product in this washed
fabric because any other contributors to formaldehyde release, 1)
unreacted agent, and 2) any autocondensation product from this
monofunctional agent, should have been removed by the washing step.

Physical evidence of crosslinking on a microstructural or
morphological level can be seen by response of cotton to methacrylate
layer expansion (44). Electron photomicrographs of cross sections of
uncrosslinked and crosslinked fibers show differences in responses to
this agent after swelling. The uncrosslinked fiber is expanded to
show the lamellae and a pore structure (Figure 5). The fiber that had
been crosslinked in a conventional manner, i.e. in the dry state,
exhibits a monolithic cross section with no lamellae separation or
visible pore structure.

The amount of moisture present at the time of crosslinking,
however, affects the behavior of the cotton fiber during methacrylate
layer expansion. With a smaller magnification (Figure 6), it can be
seen that crosslinking in a somewhat moist state permits subsequent
layer expansion, whereas the lamellae of the cotton crosslinked in the
dry state do not separate.

The amount of moisture in a cotton fabric during crosslinking
also influences the response of wrinkle recovery angle to increasing
crosslinking. The largest difference is in the response of wet
wrinkle recovery angle. Reeves, *et al.*, claimed that the level of
wrinkle recovery angle measured on fabric conditioned under ambient
conditions becomes much less than that measured on water-saturated
fabric if water content in the system at time of crosslinking is
greater than optimum (45). This phenomenon can be seen in Figure 7.

As resiliency properties, wrinkle recovery angle, recovery from
strain, and smooth-drying appearance, improve with increasing
crosslinking, the strength and toughness properties decrease because
of restriction of movement between cellulose chains. The Gulf Coast
Section, American Association of Textile Chemists and Colorists,
related the changes in fiber properties from crosslinking to changes
in fabric properties (46).

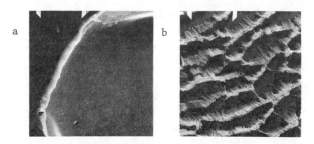

Fig. 5. Magnified cross section of cotton fibers after intrafiber polymerization of methactylate. a. Fiber crosslinked in unswollen state to give increased resiliency when dry. b. Fiber not crosslinked. (Distance between marks is 1 μ) (44)

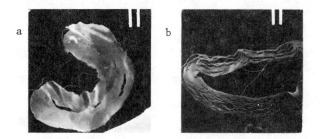

Fig. 6. Magnified cross section of cotton fibers after intrafiber polymerization of methactylate. a. Fiber crosslinked in a dry , unswollen state to give increased resiliency when dry. b. Fiber crosslinked in a swollen state to give no increase in resiliency when dry. (Distance between marks is 1 μ) (1).

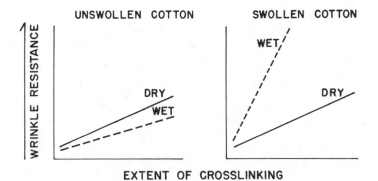

Fig. 7. Relationship between wet and dry wrinkle resistance in fabrics crosslinked in an unswollen state and in a swollen state as the extent of crosslinking is increased.

Nelson and Rousselle claim that the amount of moisture present at the time of crosslinking slows the rate of decrease in the strength and toughness properties at the higher extents of crosslinking (47). Plots in Figure 8 are from cotton fabrics given a conventional pad-dry-cure treatment (approximately 0% moisture), a mild-cure treatment and a steam-cure treatment with dimethyloldihydroxyethyleneurea (DMDHEU).

The moisture characteristics of a crosslinked cotton fabric itself vary with the amount of water present, or swelling, at the time of cure. In Figure 9 are plots of moisture regain in fabrics from room temperature treatments with formaldehyde itself as crosslinking agent (45). Moisture regain is plotted vs. extent of crosslinking in the presence of 9% water (Form D treatment) and 76% water (Form W treatment). Reduction of moisture regain by crosslinking is unchanged by the extent of crosslinking in the presence of 9% water. There is less total reduction and there is increasing moisture regain with increasing crosslinking as the amount of water is increased at the time of crosslinking.

Comparisons Between Crosslinking and Polymerization

The contrast between textile properties of cotton fabric finished predominantly with polymerization and with crosslinking is shown in Table III (48).

Table III. Effects of Deposited Polymer on Performance Characteristics of Cotton (48).

Monomer or polymer	Add-on (%)	Change in wrinkle recovery angle (degrees) conditioned (w+f)	% Change in strength related properties		
			break str.	tear str.	abrasion resist.
Methoxymethyl melamine/DMDHEU	10.0	−22	−5	−34	−68
NMP-2	7.4	+90	−45	−38	+120
DMDHEU	4.5	+92	−62	−60	−55

A fixed-only, and therefore, non-crosslinked, methylolmelamine/DMDHEU finished fabric has a high degree of polymerization, but no cellulose substitution. The fabric exhibits low wrinkle recovery, tear strength and abrasion resistance. NMP2 (N-methylol polyethyleneurea with a degree of polymerization of 2) is said to be capable, not just of linear, but also, of net-work polymerization (48), in addition to crosslinking cellulose. Fabric treated with this agent has increased wrinkle recovery and increased resistance to Stoll flex abrasion. Electron photomicrographs have been used to show increased

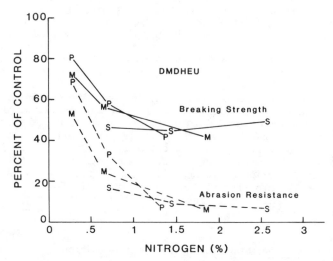

Fig. 8. Fabric breaking strengths and abrasion data expressed as percentages of control in relation to nitrogen content. P = pad-dry-cure; M = mild-cure; S = steam-cure (47).

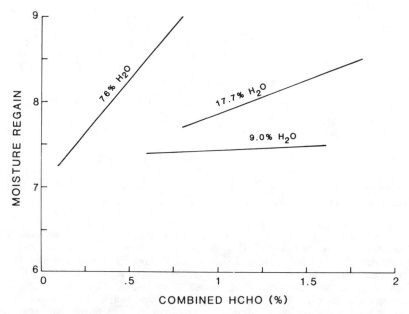

Fig. 9. Moisture regain as a result of swelling during crosslinking (45).

surface deposition of polymer with increasing time of reaction in a methylol melamine wet-fix treatment of cotton, with conditions that promote autocondensation over cellulose crosslinking (49). In Figure 10 it can be seen that as reaction time approaches 48 h at ambient temperature, the cotton fiber surface is completely obscured with obvious inter-fiber bridging.

One property affected by crosslinking to a much greater extent than by polymerization is pore size (50). Figure 11 shows how the change in pore size produced by cotton cellulose crosslinking affects Direct Red 81 dye sorption capacity. The upper three sorption isotherms are from methylol melamine/DMDHEU wet-fix treatments (WF) that have been fixed only (Figure 10). The lower three isotherms are from the same treatments that have been subjected to a curing step to effect crosslinking of the cotton by the DMDHEU (WFC). Affinity for Direct Red 81 is much reduced by the crosslinking step. A pad-dry-cure control fabric treated with DMDHEU alone had negligible affinity for the dye.

Summary

A massive amount of evidence has built up for crosslinking as the major operative mechanism in finishing of cotton for durable press. If not taken singly, certainly in combination the effects of crosslinking are convincing. There are overwhelming chemical and physical changes; the physical changes are manifested both on a gross, textile property, level, and on a microstructural, morphological level.

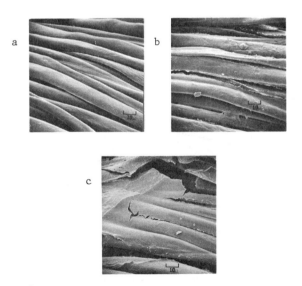

Fig. 10. Scanning electron micrographs of fibers taken from fabrics given combination polymerization-crosslinking treatments with a polymerization step of 16 h (WFC-16), 24 h (WFC-24), and 48 h (WFC-24) (49).

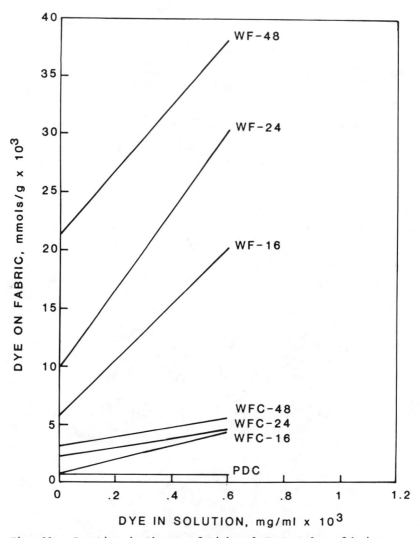

Fig. 11. Sorption isotherms of Diphenyl Fast Red on fabrics
given polymerization treatments for 48 h (WF-48), 24 h (WF-24),
and 16 h (WF-16), and combination polymerization-crosslinking
treatments with a polymerization step of 48 h (WF-48), 24 h (WF-
24), and 16 h (WF-16). PDC is a pad-cry-cure crosslinked control
(50).

Literature Cited

1. Frick,J.G.,Jr.Chem.Tech. 1971,1,100-7.
2. Meunier,L.;Guyot,R.Rev.Gin.Colloides 1929,7,53.
3. Cameron,W.G.;Morton,T.H.J.Soc.DyersColour. 1948,64(10),329-36.
4. Nickerson,R.F.Am.Dyest.Rep. 1950,39(1),P46-50.
5. Steele,R;Giddings,L.E.,Jr.Ind.Eng.Chem. 1956,48(1),110-14.
6. Foulds,R.P.;Marsh,J.T.;Wood,F.C.;Boffey,J.;Tankard,J. British Patent 291 473.
7. Gagliardi,D.D.;Nuessle,A.C.Am.Dyest.Rep. 1950,39(1),P12-19.
8. Cooke,T.F.;Dusenbury,J.H.;Kienle,R.H.;Linekin,E.E.Text.Res.J. 1954,24(12),1015-35.
9. Madan,G.L.Text.Res.J. 1974,44(12),946-47.
10. Rao,J.M.;Roberts,E.J.;Rowland, S.P.Polym.Lett. 1971,9,P647-50.
11. Frick,J.G.,Jr.;Andrews,B.A.Kottes;Reid,J.D.Text.Res.J. 1960,30(7),495-504.
12. Frick,J.G.,Jr.;Kottes,B.A.;Reid,J.D.Text.Res.J. 1959,29(4),314-22.
13. Gardon,J.L.J.Appl.Polym.Sci.1961,5(18),734-51.
14. Reeves,W.A.;Vail,S.L.;Frick,J.G.,Jr.Text.Res.J. 1962,32(5),774-80.
15. Andrews,B.A.Kottes;Arceneaux,R.L.;Frick,J.G.,Jr.Text.Res.J. 1962,32(6),489-96.
16. Willard,J.J.;Turner,R.;Schwenker,R.F.,Jr.Text.Res.J. 1965,35(5),564-74.
17. Petersen,H.,presented in part at the American Association of Textile Chemists and Colorists National Technical Conference,Philadelphia,Sept 1972.
18. Vail,S.L.;Arney,W.C.;Text.Res.J. 1971,41(4),336-44.
19. McKelvey,J.B.;Berni,R.J.;Benerito,R.R.Text.Res.J. 1964,34(12),1102-4.
20. McKelvey,J.B.;Benerito,R.R.;Berni,R.J.Text.Res.J. 1965,35(4),365-76.
21. Bullock,J.B.;Welch,C.M.;Text.Res.J. 1965,35(5),459-70.
22. Rawls,H.R.;Klein,E.;Vail,S.L.;J.Appl.Polym.Sci. 1971,15,PP.341-49.
23. Rebenfeld,L.;Weigmann,H-D.;Cotton Research Notes 1970,8,2.
24. Pai,P.S.;Petersen,H.;Reichert,M. U.S. Patent 4 207 073,1980.
25. Steele,R.J.Text.Inst.Proc. 1962,53(1),7-19.
26. Andrews,B.A.Kottes;Goynes,W.R.;Gautreaux,G.A.;Frick,J.G.,Jr. Microscope 1973,21(3),161-165.
27. Mazzeno,L.W.;Kullman,R.M.H.;Reinhardt,R.M.;Reid,J.D.; Am.Dyest.Rep. 1958,47(9),609-13.
28. PiedmontSection,AATCCAm.Dyest.Rep. 1960,49,(24),P843-55.
29. Enders,H.;Pusch,G.Am.Dyest.Rep. 1960,49(1),25-38.
30. Reid,J.D.;Frick,J.G.,Jr.;Reinhardt,R.M.;Arceneaux,R.L. Am.Dyest.Rep. 1959,48,P81-90.
31. Amer.Assoc.Text.Chem.Color."AATCC Technical Manual";1985;Vol.60.
32. Wayland,R.L.,Jr.;Smith,L.W.;Hoffman,J.H.Text.Res.J. 1981,51(4),302-6.
33. Andrews,B.A.Kottes;Harper,R.J.;Reed,J.W.;Smith,R.D. Text.Chem.Color. 1980,12(11),287-91.
34. Andrews,B.K.;Reinhardt,R.M. U.S. Patent 4 488 878,1984.
35. Kasten,M.DailyNewsRecord 1980,10(107),1.

36. Tesoro,G.C.;Oroslan,A.Text.Res.J. 1963,33(2),93-107.
37. Welch, C.M.Text.Chem.Color 1984,16(12),265-70.
38. Vail,S.L.;Murphy,P.J. U.S. Patent 3 112 156,1963.
39. North, B.J. U.S. Patent 4 284 758,1981.
40. Frick, J.G.,Jr.;Harper,R.J.Text.Res.J. 1982,52(2),141-148.
41. Roberts,E.J.;Brannan,M.A.F.;Rowland,S.P.Text.Res.J. 1970,40(3),
     237-43.
42. Roberts, E.J.;Rowland,S.P.Can.J.Chem.1970,48(9),1383-90.
43. Andrews,B.A.Kottes;Harper,R.J.,Jr.Text.Res.J. 1980,50(3),177-184.
44. Cannizzaro,A.M.;Goynes,W.R.;Rollins,M.L.;Keating,E.J.Text.Res.J.
     1970,40(12),1087-95.
45. Reeves,W.A.;Perkins,R.M.;Chance,L.H.Text.Res.J. 1960,30(3),179-
     92.
46. GulfCoastSection,AATCCAm.Dyest.Rep. 1963,52(24,37-49.
47. Nelson,M.L.;Rousselle,M.A.Text.Res.J. 1975,43(4),218-27.
48. Rowland,S.P.;Nelson,M.L.;Welch,C.M.;Hebert,J.J.Text.Res.J.
     1976,46(3),194-214.
49. Bertoniere,N.R.;Black,M.K.;Rowland,S.P.Text.Res.J.
     1978,48(11),664-71.
50. Bertoniere,N.R.;Martin,L.F.;Blouin,F.A.;Rowland,S.P.Text.Res.J.
     1972,42(12),734-40.

RECEIVED January 14, 1986

Cellulose Models for Formaldehyde Storage in Wood: Carbon-13 Nuclear Magnetic Resonance Studies

B. Meyer, K. Hermanns, and V. Baker

Chemistry Department, University of Washington, Seattle, WA 98195

13C-NMR spectra of water soluble cellulose model
compounds indicate that formaldehyde is capable of
reacting with wood cellulose functions under hot press
conditions as well as at room temperature yielding
hemiacetals. The formation of hemiacetals is
reversible, and thus constitutes a reservoir for
formaldehyde storage. Due to its affinity for water,
formaldehyde released during the manufacture of UF-resin
bonded products will be retained in the aqueous phase of
wood. Wood contains about 9 wt% of moisture. Most of
this is in the S-2 secondary cell walls that consist
mainly of wood cellulose.

Even though formaldehyde release from UF-bonded wood products has
been studied for more than 25 years, only very little is known about
how formaldehyde is stored in UF-bonded wood products. In fact, it
is not even known whether storage of formaldehyde is a physical or a
chemical process. Formaldehyde is gaseous at room temperature, but
it can polymerize forming para-formaldehyde, and it readily dissolves
in water forming methyleneglycol (2). The most likely physical
=storage process is absorption by moisture. Water is present in wood
in two forms (1): free water in the cell cavities in form of liquid
and vapor, and bound water absorbed on cellulose in the S-2 layer of
the secondary cell walls. Under standard conditions of 25°C and 50%
RH wood contains a total of 9.2 wt% water. The most likely chemical
process is the reaction of methyleneglycol with wood cellulose at the
interphase on the secondary cell surface in the S-2 layer.
There have been contradictory reports about the reaction of wood
with formaldehyde from UF-resins. At room temperature, and up to the
boiling point of water, wood absorbs only very little formaldehyde.
Thus, pine chips treated with 35 wt% formaldehyde solution for 30 min
at 160°C retain less than 0.01 wt% formaldehyde (3). Forest products
scientists generally assume that UF resins do not bond to wood (4).
However, at higher temperatures, wood absorbs formaldehyde and
irreversibly changes its physical properties. Thus, after 15 hrs of
exposure at 120°C, 7 wt% formaldehyde is retained by solid oak and
causes a 50% reduction in swelling (5-8). Since wood cellulose is

0097–6156/86/0316–0067$06.00/0
© 1986 American Chemical Society

related to cotton cellulose, it is relevant to note that textile
chemists have established extensive proof that formaldehyde can react
with cotton cellulose (9,10) and can cross-link cellulose under
textile finishing conditions, i.e. during 3-5 min exposure at 150°C.
These conditions are similar to plywood and particleboard pressing
conditions.

The purpose of this chapter is to describe exploratory 13C-NMR
studies of formaldehyde-cellulose reaction model systems. Solid
state NMR spectra are still comparatively broad and do not reveal as
much detail as solution spectra (11). Furthermore, solid state NMR
studies are still cumbersome, and since no references are available
on solid state studies of cellulose-formaldehyde interactions, we
conducted an analysis of model systems for cellulose that are water
soluble. This paper reports reactions of formaldehyde with methanol,
ethyleneglycol, some select sugars, and cellobiose.

Aqueous Formaldehyde

Formaldehyde is quantitatively absorbed in water and hydrolyzes to
yield methyleneglycol:

$$CH_2=O + H_2O = HO-CH_2-OH \qquad (1)$$

Depending upon concentration methyleneglycol polymerizes at room
temperature in aqueous solution (2) forming polymethoxy
methyleneglycol:

$$HO-CH_2-OH + HO-CH_2-OH = HO-(CH_2-O)_n-OH + H_2O \qquad (2)$$

The NMR spectrum of this system is now well established (13). The
most prominent 13C-NMR peaks are listed in Table I.

Table I.   13C-NMR Peaks of Methanol-Formaldehyde Derivatives

Compound	n	C_1	C_2	C_3	C_4
$HO-(CH_2-O)_n-OH$	1	83.1			
	2	86.6			
	3	88.9	91.6		
	4		89.2	92.1	
	5		92.3	92.5	
	6			92.7	
	7				92.9
$CH_3O-(CH_2O)_n-OH$	1	90.7			
	2	94.5			
	3	95.2		83.7	

Methanol-Formaldehyde Reaction

By far the simplest possible model system for cellulose is the
reaction of monovalent alcohols such as methanol with formaldehyde.
This system is present in aqueous phase in commercial formalin
solutions that are made by partial oxidation of methanol. These

solutions contain about 37 wt% formaldehyde and 10-12 wt% methanol
(2). The formaldehyde is present in form of a mixture of methylene
glycol and polymethoxymethyleneglycol, $HO-(CH_2-O)_n-OH$, and methoxy-
polymethoxymethyleneglycol, $CH_3-O-(CH_2-O)_n-OH$, or even dimethoxy
polymethoxymethyleneglycol, $CH_3-O-(CH_2-O)_n-O-CH_3$. These methoxy
compounds are formed by condensation:

$$CH_3OH + HO-CH_2-OH = CH_3-O-CH_2-OH + H_2O \qquad (3)$$

Methoxy compounds can also be considered as hemiacetals of the type
$R-O-CH_2-OH$. The formation and even the hydrolysis kinetics of these
compounds was studied as early as 1937 (12). Their presence enhances
the solubility of formaldehyde in water. The corresponding 13C-NMR
spectra (13) are shown in Figure 1 and the shifts are listed in Table
I. Similar spectra are obtained for higher aliphatic alcohols.

Ethylene Glycol-Formaldehyde Reaction

The 13C-NMR spectrum of the reaction of ethylene glycol, i.e.
ethanediol, with methyleneglycol is shown in Figure 2. 13C-NMR
shifts are included in Table II. It is known that, upon heating,
this system can yield methylene ether bridged rings. This reaction
is catalyzed by acids or bases. The product, dioxolane, boils at
$76^{\circ}C$. Alternatively, polyacetals are formed (14). However, studying
these mixtures under room temperature conditions we find that in
neutral solution and under our conditions the main products are
hemiacetals:

$$HO-CH_2-CH_2-OH + HO-CH_2-OH = HO-CH_2-CH_2-O-CH_2-OH + H_2O \qquad (4)$$

These compounds form rapidly at room temperature with an equilibrium
concentration depending on total and relative concentration of all
reagents. The reaction is reversible and releases formaldehyde upon
dilution. The resulting 13C-NMR shifts are shown in Figure 2 and are
included in Table II.

Table II.   13C-NMR Peaks of Aqueous Ethanediol-Formaldehyde
Derivatives

Compound	C_1	C_2	C_3	C_4	C_5	C_6
$HO-CH_2-CH_2-OH$	63.84					
Glycerol	64.0	73.5	64.0			
Erythritol	64.0	73.3	73.3	64.0		
Mannitol	64.6	72.2	70.7	70.7	72.2	64.4
Glucitol	63.8	74.3	71.0	72.6	72.5	64.2
$EG-O-CH_2-OH$	50.0					
$CH_3O-(CH_2O)_2-OH$	55.6	90.5				

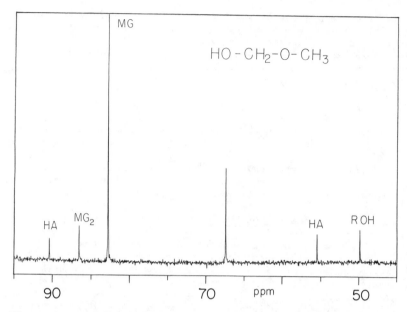

Figure 1.   13C-NMR spectrum of 1 wt% formaldehyde and methoxy
formaldehyde with 0.5 wt% methanol.   MG = methyleneglycol; HA =
hemiacetals; ROH = methanol; 67.4 = p-dioxane standard.

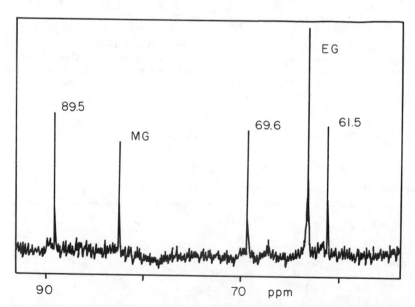

Figure 2.   13C-NMR spectrum of ethanediol-methyleneglycol
mixtures.   EG = ethyleneglycol; MG = methyleneglycol; peaks at
61.5, 69.6, and 89.5 ppm are hemiacetals.

Sugar-Formaldehyde

Pentaerythritrol, mannitol, and sorbitol react readily with
formaldehyde in the presence of zinc chloride catalyst yielding 1,2;
2,4, and 5,6 acetal bridges. Accordingly, sugar can absorb up to 5
moles of formaldehyde, but apparently not all is chemically bonded
(2). A series of authors have long noticed that evaporation of an
aqueous sugar solution containing formaldehyde yields odor free
products. It was proposed that the products might be hemiacetals
(15), but no experimental evidence was produced. The study of
interaction between sugars and formaldehyde is complicated by the
many types of products that can be formed. The literature abounds
with reports of such products, but none of these products has yet
been isolated and characterized. We have conducted exploratory
experiments with hexose compounds that were reacted with formaldehyde
(16). The 13C-NMR spectra clearly show that the products contain
hemiacetals and ether bridges, but the results are not yet conclusive
since the assignments of 13C spectra are not yet unambiguous (16,17).

Cellobiose-Formaldehyde

The structure of cellobiose and its 13C-NMR spectrum are shown in
Figure 3a. The spectra have been identified (18-20). Cellobiose is
water soluble. Figure 3b shows the spectrum of reaction products
with formaldehyde at different molar ratios obtained by 15 min
reaction at $150^{\circ}C$, i.e. under conditions that correspond to those
during the manufacture of UF-bonded wood products. As expected,
formaldehyde can react with several different functional groups.
Therefore, complex mixtures of products are formed.

Interpretation of Model Compound Reactions

In wood, as in all of the above model compounds, the formaldehyde
absorption and subsequent reaction depends on the presence of an
aqueous phase. This phase may be a monomolecular layer of water on
the cell surface, or water on the cured UF-resin film, but the
largest reservoir of water is within the wood cell. As indicated,
wood may contain two types of water: (a) free or capillary water, and
(b) bound water (1). The bound water is located in the S-2 layer of
the secondary cell walls that expand and shrink as water is absorbed
or released. The thermodynamics of the water absorption are well
established and are summarized in Figure 4. The water absorption
mechanism can be explained by two types of models. One assumes that
water forms a solution on the cellulose layer. This type of model is
exemplified by the Hailwood-Horrobin theory (12). The other assumes
water absorption on internal surfaces. This model is a modification
of the Brunauer, Emmett and Teller (BET) theory (21) that has been
expanded by Dent. Water in wood can be observed and analyzed with
proton NMR (23-25).
     When formaldehyde is released from UF resin during hot pressing
at $150-190^{\circ}C$ and elevated pressure, the vapor pressure is
sufficiently large to produce formaldehyde vapor that migrates from
the hot press platten towards the core of the product as the
temperature gradient travels to the core of the product (25). Thus,
formaldehyde permeates the product and some of it emanates from the
product edges jointly with the steam that is produced at the same

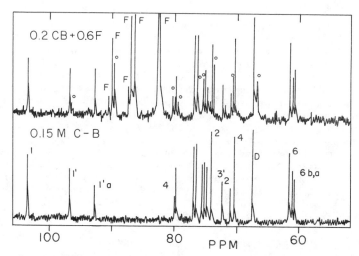

Figure 3.   13-NMR spectrum of cellobiose, (a) 0.15 M solution, and
(b) 0.2 M solution containing 0.6 M methyleneglycol.   Dotted peaks
are due to reaction products.

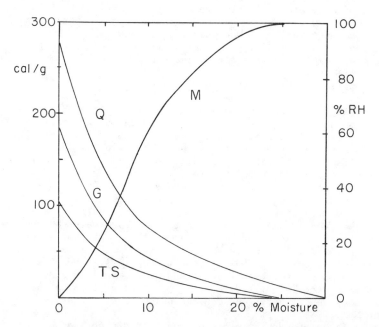

Figure 4.   Thermodynamics of water absorption on wood cellulose.
Q = heat of sorption; G = free energy; TS = entropy term; M is the
experimentally observed water sorption isotherm (after reference
32).

time. During this process formaldehyde vapor will penetrate wood
cells primarily through cell cavities, even though it is feasible for
it to penetrate the cell wall by diffusion (32). During the cooling
of the product the water content of wood may be sufficiently high to
leave temporarily some liquid water in the cell cavities, even though
cell cavities are normally dry in all except green wood. In this
case, formaldehyde, due to its propensity for water absorption,
would collect in the cell cavities. In any case, whenever
formaldehyde reaches the interior of the wood cell it will be
strongly attracted and preferentially bound in the water layer on the
surface of the S-2 cellulose layer of the secondary cell walls. This
transport of formaldehyde from UF-resin to the cellulose layer will
continue during cool-down of the product which normally takes more
than a day.

Once formaldehyde reaches the bound water layer in the S-2 cell
walls it is available for reaction with the cellulose surface. Our
exploratory experiments indicate that such reaction is indeed
expected, that it causes formation of hemiacetals, readily reaches
equilibrium, and is reversible. The concentration of formaldehyde
bound in form of hemiacetal will depend on the concentration of water
as well as that of formaldehyde. Since the water concentration
depends on relative humidity of the surrounding air, the
concentration of water in the S-2 layer, and, in turn, the
concentration of the formaldehyde solution and the hemiacetal layer
will change as a function of surrounding air humidity. The mechanism
and kinetics of this reaction follow those for other acetals (26) and
are in competition with those of UF-resin formation (27-29) steps.
The hydrolysis of the former is probably easier than that of the
latter (30), and comparable to the hydolysis of polyoxymethylene
compounds (31). The kinetics are strongly pH dependent. The pH
depends on the wood species, the buffer capacity of the resin, and
the nature of the catalyst used (27).

If we consider as an example a relative air humidity of 50% and a
temperature of $25^{o}C$, the wood moisture content would be 9.2 wt% (33).
If we further consider that the product manufacturing process leaves
about 1 wt% of the formaldehyde content of the UF resin as unreacted
formaldehyde, we obtain for particleboard or medium density
fiberboard (MDF), where UF-resin makes up 6-10 wt%, an approximate
formaldehyde concentration of 0.2 M in the S-2 cell of the wood.
This is sufficient for partial conversion to cellulose-hemiacetal,
with a residual formaldehyde concentration of less than 0.1 wt% in
the cell water. This formaldehyde concentration is enough to produce
an equilibrium vapor pressure of 20 Torr of formaldehyde (34) in the
wood cell. The kinetics of the formaldehyde release from water are
also pH dependent (35).

This vapor acts as a driving force for formaldehyde diffusion
from the wood cell towards the product surface, and for emission from
the finished wood product. An internal vapor pressure of 20 Torr
would approximately correspond to a formaldehyde air concentration of
about 1 ppm at $25^{o}C$, a load factor of 1 m^{-1} and a ventilation rate of
1 ach. However, as emission continues and depletes the methylene
glycol concentration in the wood moisture, the dissociation of
hemiacetals will set in and add to the formaldehyde source. The
bottleneck in the formaldehyde transport will be diffusion through
the product towards the product surface. This process depends on the
permeability of the product which, in turn, depends on diffusion

through the wood, and diffusion through the air gaps between wood
chips or wood layers that make up the product.
Under normal product use conditions, the air humidity and product
temperature will constantly fluctuate and pass through daily and
seasonal cycles. This will cause changes and reversals of
formaldehyde concentration gradients and formaldehyde transport
within the product. The limiting kinetic step is likely the moisture
diffusion through wood. It is well established that conditioning of
wood for reaching moisture equilibrium may take several days to a
week. Thus, real-life formaldehyde emission is not always strictly
an equilibrium process and real-life conditions are determined by
formaldehyde following water transport. An extreme example for such
a process may occur in buildings that contain particleboard, hardwood
plywood or urea-formaldehyde insulation foam (UFFI) in contact with a
wall cavity that contains improper moisture barriers. Under such
conditions sunshine can heat the wall sufficiently to cause moisture
to migrate in a daily cycle through the walls, starting in the
morning in the east and ending in the evening in the west, while
carrying formaldehyde vapor along.

Summary

Due to its affinity for water, formaldehyde will concentrate in wood
products in their water reservoirs. Since wood collects water in its
S-2 secondary wall on the surface of wood cellulose, formaldehyde
will come into contact with wood cellulose. This work shows that
formaldehyde can be expected to react with wood cellulose forming
hemiacetals. Since this reaction is reversible, these hemiacetals
constitute a temporary reservoir for formaldehyde within wood. This
fact may explain the complex formaldehyde release and absorption
properties of UF-bonded wood products.

Literature Cited

1. Skaar, C. "Wood-Water Relationships"; Adv. Chem. 1984, 207, 127-
    172.
2. Walker, J. F. "Formaldehyde"; ACS Monograph Series 159, 1964.
3. Ginzel, W. Holz Roh-Werkstoff, 1973, 31, 18-24.
4. Johns, W. ACS Symp. Proc. 1986, __, Chapter___
5. Stevens, M.; Schalk, J.; van Raemdonck, J. Int. J. Wood
    Preservation, 1979, 1,(2), 57-68.
6. Burmester, A. Holz Roh-Werkstoff, 1971, 29(3) 97-102, and (5)
    184-188.
7. Stamm, A. J. Tappi 1959, 1, 39-44.
8. Schürch, C. Forest Prod. J. 1968, 18(3), 47-53.
    Steele, R.; Giddings, Jr., L. E. Ind. Eng., 1956, 48, 110-114.
9. Cooke, T. F.; Dusenbury, J. H.; Kienle, R. H.; Linekin, E. E.
    Textile Res.J. 1954, 24(12), 1015-1035.
10. Kottes-Andrews, B., ACS Symp. Proc. 1986, __, Chapter __
11. Haw, J. F.; Maciel, G. Holzforschung 1984, 38, 327-331.
12. Löbering, J.; Fleischmann, A. Ber. 1937, 70, 1680-1683, and
    1713-1719.
13. Dankelman, W.; Daemen, J. M. H. Anal. Chem., 1976, 48, 401.
    Dankelman, W.; Daemen, J. M. H.; de Breet, J. J. Angew.
    Makromol. Chem, 1976, 54, 187.

14. Petryaev, E. P.; Gergalov, V. I.; Kalyazin, E. P.; Glushonok, G. K. Ukr. Khim. Zh., 1979, 45(9), 868-871 (C.A. 1980, 92, 40961).
15. Contardi, A.; Ciocca, B. Rend. Inst. Lombard. Sci, 1936, 69, 1057.
16. Gagnaire, D.; Mancier, D.; Vincedon, M. Org. Magnetic Resonance 1978, 11(7), 344-349.
17. De Bruyn, A.; Anteunis, M.; Verhegge, G. Bull. Soc. Chim. Belg. 1975, 84(7), 721-734.
18. Hall, L. D.; Morris, G. A.; Sukumar, S. J. Am. Chem. Soc. 1980, 102, 1745-1747.
19. Angyal, S. J.; Le Fur, R., Carbohydrate Research, 1980, 84, 201-209 and 137-146.
20. Hall, L. D.; Morris, G. A.; Sukumar, S. J. Am. Chem. Soc. 1980, 102, 1745-1747.
21. Brunauer, S.; Emmett, P. H.; Teller, E. J. Am. Chem. Soc. 1938, 60, 309-319.
22. Dent, R. W. Textile Res. J. 1977, 47(2), 145-152 and (3), 188-199.
23. Nanassy, A. J., Wood Sci, 1973, 5, 187-193.
24. Sharp, A. R.; Riggin, M. T.; Kaiser, R.; Schneider, M. H. Wood and Fiber, 1971 10(2), 74-81.
25. Hsi, E; Vogt, J; Bryant, R. G. J. Coll. Interface Science 1979, 70(2), 338-345.
26. Cordes, E. H.; Bull, H. G. Chem. Rev., 1974, 74, 581-603.
27. Glutz, B. R.; Zollinger, H. Helv. Chim. Acta. 1969, 25, 1976-1984.
28. Petersen, H. In "Chemical Processing of Fibers and Fabrics"; Marcel Dekker: New York, 1982.
29. O'Connor, C. Quart. Rev. 1970, 24(4), 553-564.
30. Vail, S. L. Text. Res. J. 1969, 39(8), 774-780.
31. Stanonis, D. J.; King, W. D.; Vail, S. L. J. Appl. Polym. Sci. 1972, 16, 1447.32.
32. Siau, J. F. "Transport Processes in Wood"; Springer-Verlag: New York, 1984.
33. "Wood Handbook," U.S. Forest Products Laboratory, Agriculture Handbook No. 72, U.S. Department of Agriculture, 1974.
34. Meyer, B. "Urea-Formaldehyde Resins"; Addison-Wesley Publishers: Reading, MA, 1979; p. 31.
35. Bell, R. P.; Evans, P. G. Proc. Royal Soc. 1966, 291A, 297-329.

RECEIVED January 14, 1986

7

Urea–Formaldehyde Resins

William E. Johns and A. K. Dunker

Wood Engineering Laboratory, Washington State University, Pullman, WA 99164-3020

Urea-formaldehyde resin solutions are shown to be dominated by physical associations rather than primary
chemical bonding. These physical associations, or
colloidal dispersions, are directly related to the
thermodynamic balance of secondary bond formation between resin and solvent systems. Steric and entripic
evaluations of molecule configuration have shown that
linear urea-formaldehyde oligomers resemble polypeptides, and have the potential to form both β-sheets
and π-helixs. While the exact configuration of the
associations is not known, their presence has been
confirmed by x-ray analysis, which shows that urea-
formaldehyde resins are crystalline in solid form.

It seems only fitting that the most commonly used resin in the world
today is based on the first organic compound to be synthesized entirely from inorganic materials. Today urea-formaldehyde (UF) resins
are produced at the rate of millions of tons per year. It is interesting that this most common of synthetic binders is one of the most
puzzling to work with and understand. This paper will review some
recent work on the nature of UF resins from a somewhat different approach; that of colloidal dispersions which are similar to another
more commonly investigated high molecular weight polymer, proteins.

Discussion

Urea-formaldehyde resin, like phenol-, or furfuryl alcohol-formalde-
hyde resins, is typically thought of as resulting from simple condensation chemistry. The ultimate hardening of the resin is thought to
be the result of the formation of a cross-linked network brought
about by acid catalysis. Current reviews are available (1, 2) which
discuss this traditional preception of UF resin chemistry.
     In many interesting ways, UF resins are different from other
types of condensation polymers. While other liquid resins are clear,
UF is typically white or cloudy. Heating a resin such as phenol-

0097–6156/86/0316–0076$06.00/0
© 1986 American Chemical Society

formaldehyde will result in a slow, predictable increase in viscosity, while a UF will remain virtually unchanged in viscosity until
gelling, at which time the resin turns almost instantaneously into
a solid.

In the formulation of resins such as phenol-formaldehyde or
epoxy resins, stoichiometric requirements call for a 2+:1 mole ratio
of reactants to achieve a high cross-link density. UF resin can be
prepared at mole ratios on the order of 1:1.10 with little problem.

During the manufacture of UF resin with a typical cook, an extended acid hold will result in a relatively high viscosity. The
addition of dry urea solids both increases the solid contents and
produces a substantial drop in viscosity.

Finally, the overall behavior of urea toward formaldehyde is
much different than is the behavior of, for example, formaldehyde
with phenol. Mixing phenol and formaldehyde at a ratio of 4:1 in
an acidic medium will result in a reaction of impressive vigor. At
ratios of 4:1 urea and formaldehyde are not capable of advancing
under acidic conditions even with the application of heat. Urea-
formaldehyde concentrate, a stable mixture of urea and formaldehyde
at a mole ratio of 4.8:1 and concentrations of as high as 85% solids,
is a common material of commerce. These observations, taken together, are not consistent with the orderly formation of a urea-formaldehyde condensation polymer.

In order to more fully explain the nature of the UF system,
Pratt and co-workers (3), investigated the potential for explaining
UF resins as colloidal dispersions rather than oligomeric solutions
and found the results most interesting. Pratt's model considered
the implications of colloidal behavior as resulting from the condensation of urea and formaldehyde to an oligomer. At some point in
the course of a typical resin cook, this oligomer would coalesce to
form a stable colloidal particle. This initial coalescing would result in the formation of the clowdy UF, typical in large scale manufacture. The concentration of formaldehyde was considered important
in the formation of this colloidal system. An excess of formaldehyde was suggested as forming a protective sheath around the UF particle and thus stabilize it. Hardening was accomplished by gradually consuming formaldehyde in continuing reactions of urea and formaldehyde with acid catalyst. At some point, it was suggested,
there would be insufficient formaldehyde to stabilize the colloidal
particle and the system would harden by coalescing.

If the hardening of a UF is simply the coalescing of a colloid,
it should be possible to see the colloidal particle in the hardened
state. Scanning electron microphotographs of hardened UF polymer
are shown in Figures 1 and 2. Figure 1 shows UF resin collected by
precipitation from a dilute solution of UF resin. Figure 2 shows a
fracture interface of a solid UF plug cured by acid catalysis.
Shown are structures of a nodular nature very similar to silica and
carbon colloids (3).

Pratt's model for the stabilizing influence of formaldehyde on
associated liquid systems is not without precedents. Terbilcox (4)
investigated the reactions of formaldehyde with calcium and ammonium
lignosulfonates under acidic conditions. An increase in viscosity
was noted with cooking for the ammonium lignosulfonate, but not the
calcium-based lignin. This viscosity increase was reported in the

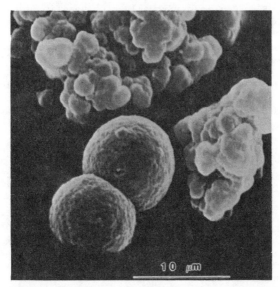

Figure 1.   Scanning electron photograph of urea-formaldehyde resin.   This specimen was prepared by the dilute solution precipitate techniaue (7).

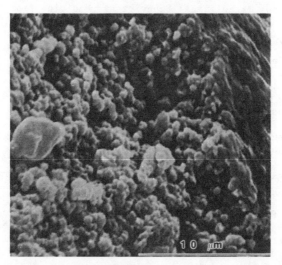

Figure 2.   Scanning electron photograph of urea-formaldehyde resin.   This surface was exposed by simple fracture of a solid plug of solid resin.

literature as resulting from the methylolation of the lignin moiety
and the corresponding increase in molecular weight of the lignin due
to subsequent condensation. When an attempt was made to determine
the molecular weight increase in the lignin via liquid chromatography, it was seen that the extended cooks did, in fact, produce
chromatograms which showed increased molecular weight.

A problem was encountered by Terbilcox when the rate of formaldehyde consumption during the cook was determined. Most of the formaldehyde disappeared immediately when mixing of the formaldehyde
and ammonium lignosulfonate occurred. This seemed unreasonable.
The chromatograms were rerun, this time with a 0.1N solution of LiCl
instead of water as the solvent system. No increase in molecular
weight was noted for any period of heating. The increasing viscosity was assigned indirectly to the consumption of formaldehyde in
the formation of hexamethylene tetramine. Here, the formaldehyde
reacted with the ammonium ion from the lignosulfonate. Formaldehyde
in the form of methylene glycol, is an excellent solvent for lignosulfonates. Its removal permits the lignin moieties to coalesce.
It should be noted that a LiCl solution is accepted in high pressure
liquid chromatography as an excellent way of disrupting the association of molecules in order to determine their true molecular weigh.

Urea–formaldehyde condensates show a surprisingly similar behavior to the lignin salts investigated by Terbilcox (4). The
ability to produce a material such as UF concentrate demonstrates
the solvent ability of hydrated formaldehyde. It is often seen that
a fresh cook of a UF is clear, and will remain so for a short period
of time. UF resins above a mole ratio of 1:2.5 (U:F) are relatively
easy to produce as a clear liquid. Typically resins which are produced at the very low F:U ratios are the most difficult to make
clear and are the least stable.

The first U.S. patent (5) on UF resin which was issued to Hanns
John, suggests that urea and formaldehyde be cooked at mole ratios
of 2:1 or 3:1 and high solid contents. The resulting product is
said to be ". . .fluid in the heated state, but it will gelatinize
when being cooled. In this way prepared, the product forms a
colorless transparent, tensile and elastic mass, insoluble in water
as well as in alcoholic solvents, and which is acted upon only by
acids, or alkali liquors." This reported ability to be heat reversible and to remain clear is similar polypeptides and agar systems.

An interesting implication of a colloidal model as suggested by
Pratt for UF resins lies in the possible structures that may result
from the hardened coalesced material. If colloidal particles do
form oligomeric UF condensates, the process of coalescing should be
ordered in a systematic way.

The work of Rammon (6) characterized UF resins prepared from UF
concentrates. One of his observations was that cured UF resins are
crystalline. Rammon's observations were subsequently confirmed by
Stuligross and Koautsky (7). This is somewhat surprising in that a
cross-linked material, by definition, is not crystalline in nature.
The cross-links should serve to disrupt the structural regularity
required to permit a crystal to form. While a study of the crystalline nature of UF resins was not the major thrust of Rammon's research, a brief survey of the phenomenon was made. Rammon showed
that all UF resins below a mole ratio of 1.43 gave distinct powder

patterns, while those above 1.43 mole ratio patterns were still dis-
cernible, but not as distinct. The nature of the resin cook did not
seem to matter, nor did the method of hardening the resin. The d-
spacing of the urea x-ray pattern fit some, but not all, of the d-
spacings of the UF resin, suggesting that some part of the UF crystal
is based on the urea molecule.

It is tempting to suggest that the organization present in the
liquid is carried over to the solid state. Rammon states the following:

> This suggests that the ordered structure is present in
> the resin solution as a liquid crystal and is maintained
> into the cured state. The presence of a liquid-crystal
> phase in natural proteins and synthetic polypeptides is well
> documented. The liquid-crystal structure is the result of
> an unique conformation which allows a highly ordered hydro-
> gen bonding system to develop. (6)

The idea of well-ordered UF structures was developed more fully
by Dunker, Johns and co-workers (8). This study compared the anticipated structure of oligomeric UF with common proteins, specifically
glycine polypeptide. The concept of UF being similar, somehow, to a
polyglycine molecule is based on two factors: the similarity of the
chemical structure and thermodynamic considerations in the solubility
of urea and formaldehyde as they condense.

Figure 3 shows the structure of glycine and a substituted urea.
To facilitate the comparison, several assumptions (8) were made.
First, oligomeric UF has minimal methylene ether linkages. This was
confirmed by Rammon (6) who studied the 13C spectra of a variety of
UF resins and found a minimal number of ether structures. Second,
similar to peptides, the N-C=O bond of urea is planer, a consequence
of the resonance of the nitrogen electrons with the carbonyl electrons as shown in Figure 4. The inability of the N-C=O bond to rotate freely has been well documented for proteins and seems reasonable to assume the same behavior for urea in light of the planarity
of the urea in crystalline form.

Figure 5 shows a schematic representation of the urea molecule
and identifies two angles ψ and ϕ . Dunker, Johns, and co-workers
showed how these two angles are limited to a specific limited range
of values. In a manner similar to that applied to polypeptide analysis, computer simulations of all possible angles based on steric
factors and configurational entropies were completed. This type of
analysis yields characteristic plots, known as Ramachandra plots. A
Ramachandra plot for UF resin is shown in Figure 6 and polyglycine in
Figure 7. Based on these computer assisted models it was possible
for Dunker, Johns and co-workers to suggest structures of the shape
of the hydrogen bonded units to the UF resin. These are shown in
Figures 8 and 9. Two types of arrangements of substituted ureas was
possible, a Π-sheet, and a β-helix.

That a UF resin should be thermodynamically capable of forming
such structures was the next problem Dunker, Johns, and co-workers
had to consider. Here the analysis was based on the effect which
methylolation has on the hydrogen-bonding balance of components and
products. Figure 10 shows the net hydrogen bond balance for a polypeptide and Figure 11 shows the net hydrogen bond balance for the
addition of two formaldehydes to a urea. The net effect of adding

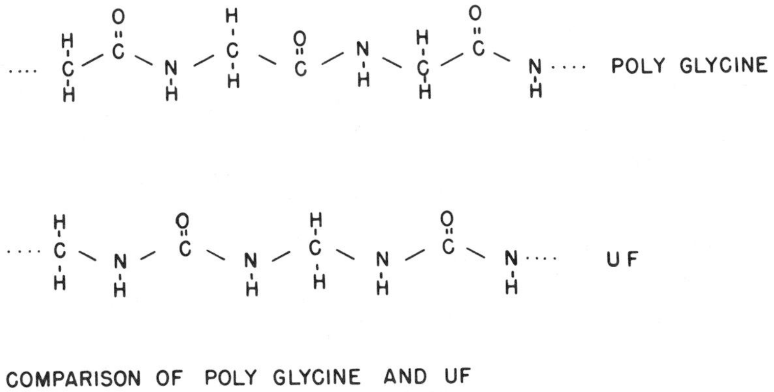

COMPARISON OF POLY GLYCINE AND UF

Figure 3. A comparison of the structures of polyglycine, a simple protein, and urea-formaldehyde resin.

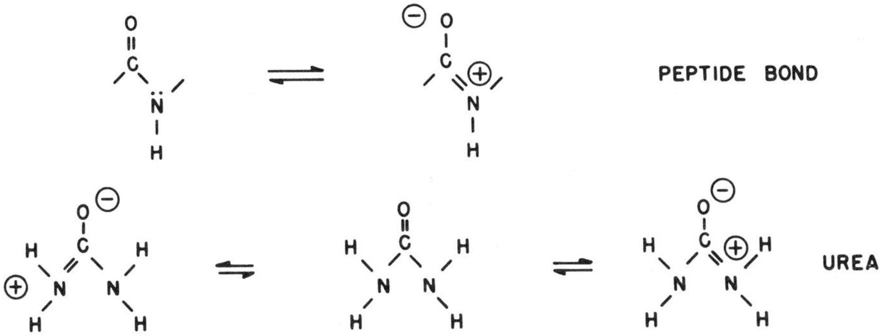

Figure 4. A comparison of the resonance of nitrogen electrons
with carbon electrons for a peptide bond common to proteins, and
the C–N bond found in urea.

82 FORMALDEHYDE RELEASE FROM WOOD PRODUCTS

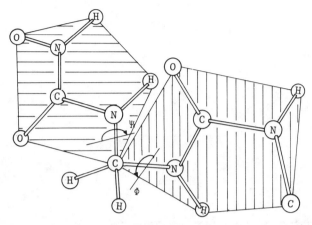

Figure 5. A schematic drawing of urea showing the two planes of
the urea molecule. The O=C-N bond of urea is not free to rotate,
while the nitrogen-methylene bridge is free to rotate. Assuming
a linear urea-formaldehyde molecule, there was two such bonds
that can rotate, here identified as ϕ and ψ.

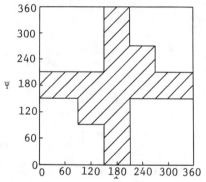

Figure 6. Ramachandran plot for urea-formaldehyde resin. The
cross-hatched area identifies forbidden angles for ϕ and ψ.

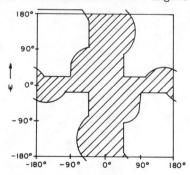

Figure 7. Ramachandran plot for polyglycine. The cross-hatched
area identifies forbidden angles for ϕ and ψ.

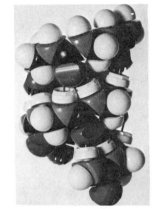

Figure 8.   End view (a) and side view (b) of a Ⅱ helix.   This is a proposed model for a urea-formaldehyde resin based on colloidal considerations.

Figure 9.   Side view of a β-sheet.   This is a proposed model for a urea-formaldehyde resin based on colloidal considerations.

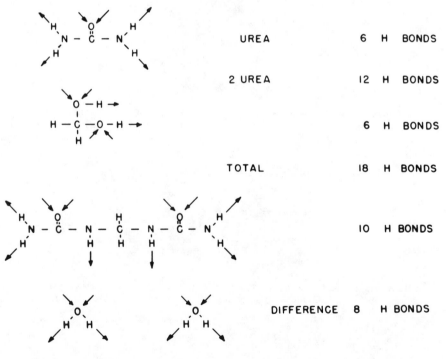

	GLYCINE	7 H BONDS
	2 GLYCINES	14 H BONDS
	GLYCLGLYCINE	10 H BONDS
	DIFFERENCE	4 H BONDS

Figure 10. A balance diagram for the condensation of glycine.
The net difference is 4 hygrogen bonds.

	UREA	6 H BONDS
	2 UREA	12 H BONDS
		6 H BONDS
	TOTAL	18 H BONDS
		10 H BONDS
	DIFFERENCE	8 H BONDS

Figure 11. A balance diagram for the condensation of two methylene glycol molecules with one urea molecule. The net difference
is 8 hydrogen bonds.

two methylene glycol units to one urea is to decrease the enthalpic
contribution to solubility by 12-24 kcal/mole. This, based on very
conservative estimates, corresponds to a decrease in solubility by a
factor of approximately 10^4. Thus, the simple formation of oligomeric moieties leads to a dramatic decrease in solubility. The conclusion reported by Dunker, Johns, and co-workers is that if the organic
unit is not capable of forming strong hydrogen bonds with water,
there is a thermodynamic predisposition to develop inter- and intramolecular hydrogen bonds ultimately leading to the formation of the
colloidal dispersion.

Changes in the hydrophilicity of UF resins are not uncommon.
During an acid advance of a UF cook, solid urea is commonly added.
This both adjusts the mole ratio of the cook to the desired level,
and has the added advantage of reducing the viscosity. A surprisingly small amount of water will have the same effect on viscosity.
Here, the amount of water added will not be sufficient to act as a
diluting factor. Since the UF polymer becomes more hydrophobic with
increasing molecular weight, the presence of additional water tends
to drive the UF colloid more to intramolecular bonds and fewer UF-
water bonds. Thus, the formation of colloidal associations. When
taken to its logical extreme, the addition of an excess amount of
water will cause the UF resin to precipitate, which has been noted
(3, 8).

The implications of this work on the understanding and control
of formaldehyde release from UF systems are significant. In a model
of UF condensates, Pratt (3) suggested that formaldehyde is involved
in the formation of a protective sheath surrounding particles of UF
condensate. This protective sheath provides stability of the UF colloid; the failure of the protective sheath of formaldehyde leads to
hardening. It is commonly known that during the cure of a UF, there
is a large formaldehyde release, far greater, for instance, than with
the cure of a comparable amount of phenol-formaldehyde resin. These
observations directly lead to the speculation that if all the formaldehyde in a UF resin could be involved chemically rather than just
physically, a UF polymer of increased properties and lower emissions
could be made.

Summary

This paper has attempted to show recent observations on the nature of
UF resins. It is not comprehensive since, at the time of the Symposium, little has been firmly proven. Yet, the implications of the
research reported here are significant. The application of techniques similar to those used in the field of bilchemistry lend themselves to the illumination of the structure and behavior of common
wood resins. Also to be noted, is the possible importance of the
physio-chemical rather than the chemical qualities of a wood binder.
Finally, the net quality of the UF resin is now in a position to be
considered more carefully. If one considering UF technology only
from the perspective of organic chemistry, few major improvements
beyond the lowering of the U:F ratio with the corresponding reduction
in formaldehyde emissions have been realized recently. If the suggested model for UF resins is correct, then perhaps there is much to
be gained by enhancing the solubility of UF condensates so as to

permit the UF resin to complete the chemical reactions which the
coalesced colloid tend to inhibit.

Acknowledgments

I wish to thank Dr. Tom Pratt for valuable discussions relating to
the nature of urea-formaldehyde systems.

Literature Cited

1. Meyer, Beat. Urea-formaldehyde Resins. Addison-Wesley Publish-
    inc Co., Inc. 1979
2. Pizzi, A. Aminores in Wood Adhesives, chapter 2 in Wood Adhe-
    sives: Chemistry and Technology ed. by A. Pizzi, Marcel Dekker,
    N.Y. 1983.
3. Pratt, T.J., Johns, W.E., Rammon, R.M., Plagemann, W.L. 1984.
    A novel concept of the structure of cured urea-formaldehyde
    resin. J. of Adhesion. Vol 17(x), page xxx.
4. Terbilcox, T.F., 1983. Formaldehyde modified lignosulfonate
    extenders for furan systems. M.S. thesis, College of Engineering,
    Washington State University, Pullman, Washington.
5. John, H. October 19, 1920. Manufacture of aldehyde condensation
    product capable of technical utilization. U.S. Patent No. 1,
    355,834.
6. Rammon, R.M. 1984. The Influence of Synthesis Parameters on
    the Structure of Urea-Formaldehyde Resins. Ph.D. Thesis,
    Washington State University, Pullman, Washington.
7. Dunker, A.K., Johns, W.E., Rammon, R.M., Farmer, B., Johns, S.J.
    Slightly bizarre protein chemistry: urea-formaldehyde resin
    chemistry from a biochemical perspective. Submitted to the
    Journal of Adhesion.

RECEIVED January 14, 1986

Mechanisms of Formaldehyde Release from Bonded Wood Products

George E. Myers

U.S. Department of Agriculture, Forest Products Laboratory, One Gifford Pinchot Drive, Madison, WI 53705-2398

Published studies on wood systems and my recent research on the influence of urea-formaldehyde (UF) resin hydrolysis on formaldehyde emission from UF-bonded wood products indicate that (a) in an acid-catalyzed UF board, formaldehyde can exist in a wide variety of states, including dissolved methylene glycol monomer and oligomers, paraform, hexa, chemically bonded UF resin states, chemically bonded UF-wood states, cellulose hemiformals and formals. Each of those states is a potential source of formaldehyde emission by evaporation (methylene glycol) or initial hydrolysis. We cannot now quantify the relative contributions of these states over time; (b) in a base-catalyzed phenol-formaldehyde (PF) board, formaldehyde states may include methylene glycol monomer and oligomer, chemically bonded PF resin states, chemically bonded PF-wood states, cellulose hemiformals. Emission sources apparently include methylene glycol, cellulose hemiformals, and possibly phenolic methylols; and (c) diffusion processes very likely exert a major influence on panel emission rates and may involve movement of methylene glycol in the wood's moisture or of gaseous formaldehyde within the board or within the board-air interface.

Over the past decade or so, great progress has been made in reducing formaldehyde emission from wood products such as particleboard, hardwood plywood paneling, and medium density fiberboard (1-3). Beneficial steps include reducing the formaldehyde-to-urea (F/U) mole ratio (4), impregnating the wood furnish (substrate) with a formaldehyde scavenger having hindered access to the urea-formaldehyde (UF) adhesive (5), and treating boards with formaldehyde scavengers and/or barrier coatings after manufacture (6). Many plants in Europe now produce particleboard, for example, that meets the German E-1 standard recommending large test chamber formaldehyde levels of

<0.1 ppm (7). The United States wood products industry is now pro-
ducing particleboard and hardwood plywood paneling that meet the
recently imposed Housing and Urban Development (HUD) product
standards aimed at maintaining formaldehyde levels in new mobile
homes < 0.4 ppm (8).

Despite this practical progress, great uncertainty still exists
as to the precise mechanism by which formaldehyde is held within a
board and slowly released as a gas to the atmosphere. Historically,
many have considered the emission potential of a board to be governed, particularly in a board's early life, by the board's so-called
"free" formaldehyde content (9).

This "free" formaldehyde is presumed to derive from excess formaldehyde present in the UF resin. It exists in ill-defined, relatively loosely bound states within the board, states whose stabili-
ties are sensitive to temperature and humidity. At high resin F/U
ratios, the "free" formaldehyde content and board emission rate fall
rapidly after pressing and later decrease more slowly. The "free"
formaldehyde content and board emission rate are lower after pressing
when using resins with F/U ratios approaching 1.0, and they decrease
more slowly with time. What has never been clear, however, is
whether actual UF resin hydrolysis, with attendant formaldehyde
production, is responsible for a significant amount of the board's
emission, and if so, at what point in the board's life that occurs.

The question of the contribution of UF resin hydrolysis to board
emission is not a trivial one. If resin hydrolysis contributes significantly to emission, then, in principle, the board would retain
the potential to emit during its useful life, in contrast to the
situation if all the emission results from "free" formaldehyde. In
the former case, efforts to minimize emission must be directed toward
resin stabilization and/or to ensuring that incorporated formaldehyde
scavengers retain their effectiveness at low formaldehyde activities
for the board's entire useful life. Another consequence of continued
resin hydrolysis is possible limits on the durability of UF bonded
products; in this case improvement may be expected from more stable
resins.

Objective and Approach of Paper

The overall objective of this and a companion paper (10) is to define
the extent to which board formaldehyde emission is controlled by
resin hydrolysis or other processes. In the companion paper I have
critically reviewed the literature and presented original Forest
Products Laboratory (FPL) data in three related aspects of the
formaldehyde emission phenomenon: the chemistry of and formaldehyde
liberation from formaldehyde-urea and formaldehyde-phenol states; the
chemistry of and formaldehyde liberation from formaldehyde-cellulose
and resin-cellulose states; and our knowledge of the board emission
mechanism derived from actual board and wood systems. Whereas my
oral presentation at the American Chemical Society (ACS) Symposium
made use of information from all three of those parts, this written
paper, in the interest of saving space, is limited to literature and
FPL data dealing with actual wood-containing systems. The Conclusions section of this paper, however, makes use of the results from
all three parts of the companion paper. Experimental details of the

recent FPL testing are in the Appendix, as are explanations of calculation procedures.

Factors to be considered in this paper include (a) the degree
to which formaldehyde emission rate from wood systems is controlled
by diffusion processes, (b) the contribution of resin hydrolysis to
emission rate, and (c) the contribution of formaldehyde-wood states
to emission rate. In the following, therefore, I first summarize
briefly the reported evidence regarding diffusion control and resin
hydrolysis in actual bonded wood products. Thereafter, I present and
discuss some of my own recent experiments on wood systems that
attempted to shed additional light on the questions of resin hydrolysis and the emission mechanism more generally.

Literature Evidence for Diffusion Control

Although published evidence is sparse, there is little doubt that
diffusion processes can play an important role in board emission.
Some of the more critical findings are as follows:

(a) Particleboard emits two to three times less formaldehyde
after conditioning than do exposed core surfaces (11).

(b) Emissions are higher from board edges than from board faces
(several studies, including 12).

(c) Emission levels are decreased at higher board density
(12,13) and at lower board porosity (12).

(d) Ventilation rate and board loading effects on emission
levels in chambers can be quantitatively described by equations that
are based upon the assumption that diffusion across a board-air
interface layer governs the emission rate (14). At sufficiently high
ventilation rates, the dependence on ventilation rate disappears and
formaldehyde loss is governed by within-board processes (15).

It appears, therefore, that formaldehyde emission rate from a
given large panel may be controlled by chemical processes within the
board or by diffusion either in the board-air interface or within the
board. Which of these predominates depends upon the board's age,
composition, physical structure, and exposure conditions.

Literature Evidence for Resin Hydrolysis in Actual Boards

Despite the rather massive literature on formaldehyde emission from
UF-bonded wood products, evidence for a direct causal relationship
between resin hydrolysis and formaldehyde emission from bonded
products is almost nonexistent. Indeed, evidence in the literature
that UF resin hydrolysis actually does occur in a board arises primarily from studies into the question of whether the limited durability of UF-bonded wood products is caused by resin hydrolysis or
by a particular susceptibility of UF resin-wood bonds to rupture from
swelling/shrinkage stresses.

Evidence for Resin Hydrolysis. That UF resin hydrolysis can occur in
boards is strongly indicated by the following:

(a) greater rates of strength loss for UF boards and joints
compared to those made with other adhesives (phenolics, isocyanates,
melamines) during aging at constant temperature/humidity, particularly at high temperature/humidity (16,17,18).

(b) decrease in board modulus of rupture (MOR) but not in internal bond after spraying just the surface mat with water prior to pressing (18).

(c) increase in solubility of cured resin in both UF-bonded particleboard and UF-bonded Perlite (nonswelling volcanic glass) board during aging (19).

(d) decreased strength losses during constant temperature/ humidity aging of plywood after soaking in $NaHCO_3$ to neutralize the acid cure catalyst, which would otherwise catalyze resin hydrolysis (20).

Evidence for Swelling/Shrinkage. Evidence that the lower durability of UF-bonded products can also be brought about by swelling/shrinkage stresses in a board includes the following:

(a) faster strength losses for UF boards than for others (phenolic, isocyanate, melamine) during cyclic humidity/temperature aging, where swelling/shrinkage stresses can be strong (19,21-26).

(b) greater internal bond (IB) loss and thickness swelling increase with UF particleboard than with a UF Perlite (nonswelling volcanic glass) board (19,27).

(c) no change in modulus or strength of cured neat UF resin films during humidity cycling, i.e., when no swelling/shrinking substrate is present (28).

(d) increase in thickness swelling of boards with low F/U resins both before and after cyclic weathering (29), accompanied by the postulate (28) that low F/U resins are more brittle than high F/U resins.

(e) decreased strength loss on boiling plywood bonded with UF resins containing polyfunctional ureas which are postulated to produce more flexible binder networks (30).

(f) accelerated aging under stress of UF joints relative to PF joints (31).

Ambiguous Evidence. Finally, several studies have yielded results whose interpretation is less clear-cut:

(a) far greater cumulative amounts of formaldehyde emitted by boards than can be accounted for by their Perforator (see Appendix 1c) values (32), which have often been presumed to measure primarily non-resin formaldehyde. Unfortunately, it will be shown later that the Perforator value does not necessarily measure all formaldehyde-wood states or only non-resin formaldehyde.

(b) reduced rate of cured resin film cracking by incorporating acid reactive filler. Such materials will decrease the acidity within the resin, thereby decreasing hydrolysis; however they may also reduce the extent of resin cure, thereby decreasing brittleness and tendency to crack (33).

(c) decreased strength loss of UF particleboards by using less acidic cure catalyst (18) or by incorporating acid scavengers (34), arguments here being identical to those immediately above.

(d) greater mat moisture content (MC) yielded greater formaldehyde emission during (35) and after (36) particleboard pressing. Plausible alternatives to resin hydrolysis, however, are that greater mat MC facilitates formaldehyde movement to the board surface and/or that it enhances hydrolysis of cellulose formals and hemiformals.

Overall, therefore, the available literature supports the
generally held view that the durability of UF-bonded wood products
is governed by the susceptibility of cured UF resin bonds to scission
by both hydrolysis and swell/shrink stresses. Note, moreover, that
in either case, the most likely product of scission will ultimately
be formaldehyde and further that mechanical stress enhances the rates
of many chemical reactions (37). In fact, simplistic calculations
based on formaldehyde liberated from bond ruptures at least indicate
the possibility that formaldehyde from swell/shrink stress rupture
could contribute significantly to total emission. Assume, for
example, that board failure occurs due to rupture of one chemical
bond type which liberates one molecule of formaldehyde and consider
two cases: (a) a conservative one in which only 5 percent of those
bonds rupture in 50 years, i.e., probable board durability greater
that 50 years, and (b) a much less conservative case in which 30 percent of those bonds rupture in 20 years, i.e., probably failure in
20 years or less. Case (a) leads to a first order scission rate
constant of 3.3 x 10^{-11} s^{-1} and a hypothetical board emission rate
(see Appendix 3a) that is below the maximum liberation rate permitted by the German E-1 standard (7). However, Case (b) leads to a
first order scission rate constant of 5.7 x 10^{-10} s^{-1} and a hypothetical board emission rate above that allowed by the HUD standard
(8). (Formaldehyde-wood interactions and diffusion effects would
undoubtedly lower the board emission rates from these hypothetical
values.)
      On this basis, therefore, we might expect UF resin bond scission
to be one source of board formaldehyde emission. However, the
available studies do not permit quantitative statements about the
relative magnitudes of that source compared to other sources, such as
formaldehyde-wood states, during board lifetime.

Recent FPL Studies

To shed additional light on the emission mechanism and the contribution of resin hydrolysis to formaldehyde emission, my recent experiments have examined the liberation or extraction of formaldehyde from
particleboards, from wood containing sorbed formaldehyde, and from
cured resins. Here, I present results from particleboard and formaldehyde-sorbed wood experiments in which rates of formaldehyde removal
were measured by three different procedures (see Appendix 1 for
experimental details).

Formaldehyde Removal By Gas Elution. These experiments involve the
continuous collection of formaldehyde removed by a controlled flow of
gas over the wood samples. Variables studied include time, gas flow
rate, sample comminution, gas type, humidity, and adhesive type.

      Comminution and Flow Rate Effects on Gas Elution. Elution rates
were measured from UF particleboard at two geometries--i.e., shredded
(85 pct < 1 mm) and 25x25x16 mm pieces. Shredding was conducted in a
sealed system so that no formaldehyde was lost during that operation.
The eluting gas was nitrogen at zero and 20 percent relative humidity
(RH) and at flow rates corresponding to 0.4 to 4.5 changes in gas
volume per minute (NCM).

Small effects of flow rate were found with dry nitrogen between
0.5 and 1.0 NCM but none with 20 percent RH, nitrogen between 0.8,
and 4.5 NCM. Figure 1 compares results for pieces and shredded particleboards at two levels of Perforator (see Appendix 1c) values.
Several points should be noted:

(a) Elution from shredded UF board is only slightly faster than
from the 25x25 mm pieces, and the increase is consistent with
observed effects of the flow rate difference (1.0 NCM for shredded
versus 0.5 for pieces). This similarity in elution rates indicates
that the rate-limiting step in formaldehyde release in these experiments is not "macro-diffusion" within voids but is either "micro-
diffusion" within the wood or an actual bond rupture step.

(b) In none of the tests on shredded UF board is any burst of
liberated formaldehyde observed during shredding of the 25x25x16 mm
pieces. Apparently, no significant amount of formaldehyde exists as
gas within voids, i.e., all formaldehyde in the board pieces is
present in a physically dissolved or sorbed state or in a chemically
reacted state. This is, of course, consistent with the point above
and with the high reactivity of formaldehyde with water, urea, and
wood components (10).

(c) The elution process is quite slow and has not reached any
obvious endpoint after 10 days, although the evolved formaldehyde
totals only about 20 to 30 percent of that removed by the 2-hour
toluene boiling in the Perforator test. Obviously, therefore, dry
nitrogen does not readily remove formaldehyde--caused, no doubt, by
the nonpolar nature of nitrogen and by removal of water from the
board.

Eluant Gas Effects on Gas Elution. Very brief tests were made
to compare the effectiveness of dry N_2, CO and CO_2 as eluants
(Figure 2). The three gases provided no differentiation between
formaldehyde states in UF board.

Gas Moisture Effects on Gas Elution. As expected, the influence
of moisture in the eluting nitrogen is very strong (Figures 3 and 4).
Points to be noted here are as follows:

(a) The observed absence of an endpoint to the dry gas elution
from UF board after 10 days (Figure 1) is here extended to 40 days
(Figure 4).

(b) During about 15 days of elution at 80 percent RH (Figure 3),
the UF board sample loses an amount of formaldehyde equal to approximately 80 percent of the original Perforator value and the rate shows
no indication of slowing. Similarly, at 20 percent RH a UF particleboard loses formaldehyde to the extent of about 50 percent of the
Perforator value in 40 days. (Perforator values for one UF board
were not increased by extending the toluene reflux time beyond the
standard 2 hours.) Clearly, moisture in the eluting gas removes
formaldehyde from states within the board that are not affected by
the Perforator conditions (toluene reflux, 2 hours). Whether those
states include formaldehyde bonded to resin, i.e., whether resin
hydrolysis occurs under the elution conditions, cannot be firmly
stated. However, the rapid liberation rate observed (10) for cured
resin at high humidity provides strong, indirect evidence for resin

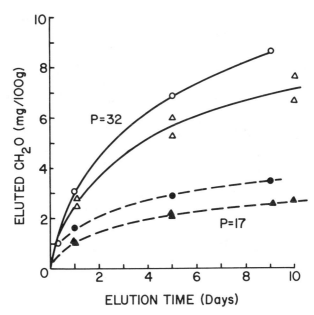

Figure 1. Particleboard elution by dry nitrogen; sample geometry effects (o ● shredded 1.0 NCM. △ ▲ 25x25x16 mm 0.5 NCM; duplicate runs. P = Perforator value in mg/100 g dry board, measured on starting material at ~6 pct moisture content.) (ML85 5428)

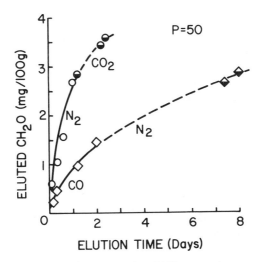

Figure 2. Particleboard elution by different dry gases. (Differences in the two curves due to different flow rates and experimental configurations. P as in Figure 1.) (ML85 5429)

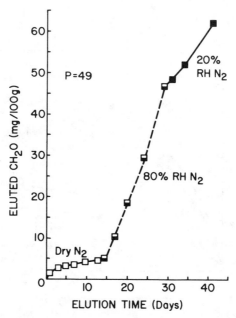

Figure 3. Urea-formaldehyde particleboard elution by nitrogen;
relative humidity (RH) effects. (0.4 NCM. P as in Figure 1.)
(ML85 5430)

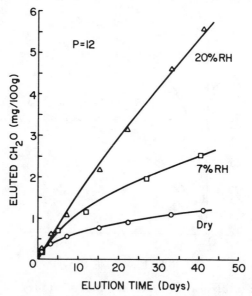

Figure 4. Urea-formaldehyde particleboard elution by nitrogen
at different relative humidities (RH). (0.5 NCM. P as in
Figure 1.) (ML85 5431)

hydrolysis contributions to the observed board losses at high humidities.

Resin Effects on Gas Elution. Elution experiments were also performed on PF-bonded particleboard and on Southern pine chips (furnish without resin) that had sorbed formaldehyde via room temperature vapor phase equilibration (see Appendix 1d and 2). Points to be noted here are as follows:

(a) the elution patterns from zero to 20 percent RH for the phenol-formaldehyde (PF) board (Figure 5) are very similar to those for the UF board. However, the formaldehyde losses for the PF board are approximately ten-fold less than for the UF, and the PF losses at 20 percent RH are likely to exceed the Perforator value sooner than in the case of the UF board.

(b) the elution patterns from zero to 20 percent RH for the formaldehyde-sorbed furnish (Figure 6) are again similar to those for the two board types, although elution rates are faster, relative to the respective Perforator values, for the furnish than for the boards. (Negligible amounts of formaldehyde were eluted from the same furnish unexposed to formaldehyde.) Obviously, the Perforator test does not measure the total of all possible formaldehyde non-resin states, even where those states are formed in the absence of heat or resin cure catalysts (furnish pH = 3.9).

Formaldehyde Liberated in Weighing Bottle Test. This test measures the formaldehyde transferred from a ground sample to a sulfuric acid solution via the vapor phase in a closed container, the acid acting as both humidity controller and formaldehyde sink (see Appendix 1a). Measurements were conducted on ground UF and PF particleboards. They were also done on ground Southern pine that had first been impregnated with tartaric acid solutions at pH 2 or 3, then vapor-sorbed with formaldehyde, and finally either aged at room temperature for 2 weeks or heated 4 minutes at 160°C to model board pressing conditions. Liberation tests were run at 27°C and at both 33 percent and 80 percent RH on -80 mesh (< 180 μm) materials and on several particle sizes between 180 μm and 62 μm. Points to be noted are as follows:

(a) At 33 percent RH (Figure 7) the formaldehyde-sorbed wood virtually completes its loss of formaldehyde after about 15 to 20 days, whereas the UF particleboard appears to be still liberating formaldehyde slowly. (The PF particleboard liberation is an order of magnitude below that of the UF particleboard and possesses poor accuracy.) Heating the formaldehyde-sorbed wood has caused either a loss of formaldehyde or a stronger bonding to the wood (perhaps formals). Liberated amounts for the formaldehyde sorbed wood equal or slightly exceed the Perforator values, while the UF board Perforator is exceeded quite early.

(b) At 80 percent RH (Figure 8) the above differences between UF particleboard and formaldehyde-sorbed wood are magnified. Liberation from the UF particleboard continues rapidly at 30 days while that from formaldehyde-sorbed wood becomes nearly constant in only about 5 days. The wood samples also liberate total amounts that are close to their Perforator values measured at high moisture. Most of the

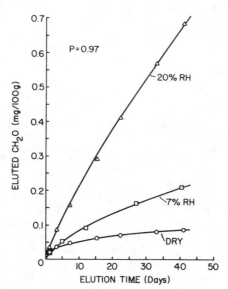

Figure 5.   Phenol-formaldehyde particleboard elution by nitrogen
at different relative humidities (RH).   (0.5 NCM.   P as in
Figure 1.)   (ML85 5432)

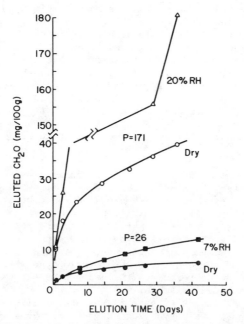

Figure 6.   Elution of formaldehyde-sorbed furnish by nitrogen at
different relative humidities (RH).   (0.5 NCM.   P as in
Figure 1.)   (ML85 5433)

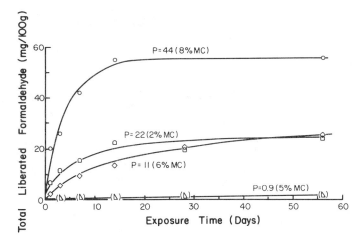

Figure 7.   Formaldehyde liberation from particleboards and
CH$_2$O-sorbed wood at 27°C and 33 percent relative humidity (RH);
weighing bottle test with -80 mesh materials (o Southern pine
impregnated with pH 2 tartaric acid and vapor-equilibrated with
CH$_2$O/salt solution at ~50 pct RH; Ο as before except heated
4 min. 160°C after CH$_2$O sorption; ◇ urea-formaldehyde particle
board; (△) phenol-formaldehyde particleboard, values approximate;
P = Perforator value at indicated moisture content (MC)).
(ML85 5434)

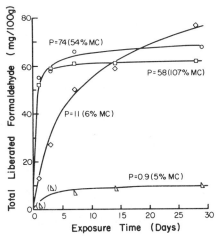

Figure 8.   Formaldehyde liberation from particleboards and
CH$_2$O-sorbed wood at 27°C and 80 percent relative humidity (RH);
weighing bottle test with -80 mesh materials.   (o Southern pine
impregnated with pH 2 tartaric acid and vapor-equilibrated with
CH$_2$O/salt solution at ~75 pct RH; □ as before except pH 3
tartaric acid; ◇ urea-formaldehyde particleboard; △ phenol-
formaldehyde particleboard, parentheses indicating approximate
values; P and MC as in Figure 7.)   (ML85 5435)

formaldehyde in the formaldehyde-sorbed wood is, therefore, very
weakly bonded (perhaps hemiformal and methylene glycol) although
there may be small quantities that are liberated with greater difficulty, particularly at pH 3 relative to pH 2. The UF board, in contrast, apparently contains little of the very loosely bound formaldehyde but contains greater amounts of more strongly bound formaldehyde, as would be expected. The PF board liberation is again well
below that of the UF board and behaves similarly to the formaldehyde-
sorbed wood samples except for greatly exceeding its Perforator
value.

    (c) At 80 percent RH the UF board exhibits no significant par-
ticle size effects on liberation rates between particle sizes of
approximately 60 and 180 µm. In that size range, therefore, within-
particle diffusion does not influence liberation rate from the UF
board.

Formaldehyde extracted in water. Formaldehyde liberated during continuous exposure to water at pH 3 was also measured on the same materials as employed in the weighing bottle test. Very dilute slurries
of -80 mesh material were held at 25°C in the presence of sodium
azide as bacterial inhibitor (Appendix 1e). In 1 or 2 hours almost
all removable formaldehyde is extracted from the formaldehyde-sorbed
wood samples (Figure 9), the total amounts being nearly identical to
those liberated at 80 percent RH and to the Perforator values. However, liberation from the UF board continues rapidly after 6 days and
at 30 days far exceeds the amounts at 80 percent RH and the amounts
from the wood samples in water. Interestingly, liberation from the
PF board in water also exceeds that at 80 percent RH and may be
occurring in two or more stages; even the apparent initial stage,
however, is an order of magnitude greater than the Perforator value.

Interpretation and Extrapolation
  to Boards in Service

In this section, I offer an analysis of these experimental results
and speculate about their implications for large panel formaldehyde
emission.

Interpretation for Comminuted Systems. The similarities and differences noted for the kinetics of formaldehyde removal from UF and PF
particleboards and from formaldehyde-sorbed wood are brought out more
clearly by plotting relative formaldehyde losses versus time. Loss
ratios, i.e., formaldehyde loss by any material divided by the UF
board loss at the same time, are shown in Figures 10 and 11; included
in Figure 10 are analogous ratios for resin data from formaldehyde
liberation (weighing bottle test) and formaldehyde elution by toluene
experiments (10). Examination of the data leads to the following
additional comments:

    (a) Southern pine containing formaldehyde that was sorbed at the
wood's natural pH or at pH 2 to 3 holds the formaldehyde in a state
that is strongly retained at low humidity but relatively labile at
moderate to high humidities. The formaldehyde is nearly completely
released, for example, in 12 days at 33 percent RH (Figure 7), in
5 days at 80 percent RH (Figure 8), and in 0.2 days in pH 3 water

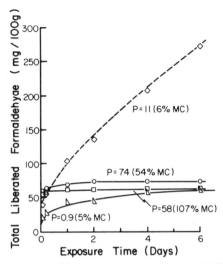

Figure 9. Formaldehyde liberation in water at 25°C and pH 3 from particleboard and CH_2O-sorbed wood; all materials -80 mesh. (Sodium azide in water at 100 mg/L as preservative; symbols and abbreviations as in Figure 7.) (ML85 5436)

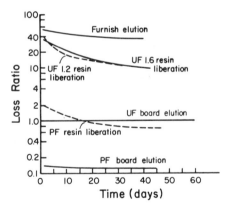

Figure 10. Formaldehyde loss ratios at 20 percent relative humidity for various materials. (Formaldehyde removed from a material divided by that removed from urea-formaldehyde particleboard. Board elution by nitrogen. Resin liberation by weighing bottle test. PF = phenol-formaldehyde) (ML85 5437)

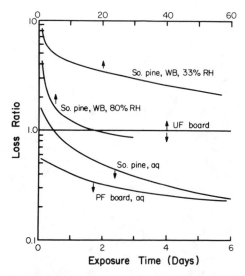

Figure 11. Formaldehyde loss ratios at 80 percent relative
humidity (RH) and in water. (Loss ratio = CH_2O liberated relative to that from urea-formaldehyde particleboard in same test.
WB = weighing bottle test; PF = phenol-formaldehyde; aq = water
extraction test at pH 3. All materials -80 mesh. Southern (So.)
pine impregnated with pH 2 tartaric acid and CH_2O vapor-sorbed.)
(ML85 5438)

(Figure 9). The available information (10) indicates this formaldehyde is present as monomer (methylene glycol) or oligomer dissolved in the wood's moisture, or possibly as cellulose hemiformals.

(b) With PF board the amount of formaldehyde released within the time scale of these experiments varies more greatly with humidity and does not as obviously exist in only one state (cf. Figures 4,7-9). While a portion of the removable formaldehyde very likely exists in the same state(s) as in the formaldehyde-sorbed wood, a major portion is more strongly held but still sensitive to moisture. The latter state perhaps is phenolic methylols (10).

(c) The UF board undoubtedly contains some of the same moisture-labile states that are present in the formaldehyde-sorbed wood, and these account for some of the initially rapid loss observed at 80 percent RH (Figure 8) and in water (Figure 9). Up to 20 percent RH the release pattern from the UF board by nitrogen elution is very similar to that from the PF board and formaldehyde-sorbed wood, indicating similar release mechanisms from all three comminuted wood systems under those conditions (Figure 10). In the other types of experiment at higher humidities, however, the release pattern from the comminuted UF board clearly differs from those in the other two wood systems (Figure 11). The continued evolution of formaldehyde from the UF board beyond the very early portion and at rates increasing with humidity strongly indicates extensive hydrolytic sources other than those present in the PF and formaldehyde-sorbed wood. Obviously, those additional sources are most likely UF resin and UF-wood states, with some possibility of cellulose formals (10).

(d) Point (c) suggests a similar release mechanism for the shredded boards and furnish particles during nitrogen elution at 20 percent RH and below (Figure 10). This implies identical rate-limiting steps, which might be a chemical bond rupture or a monomeric formaldehyde diffusion process. If that step is chemical, the nature of the three systems dictates that it most probably involves hydrolysis of cellulose hemiformals (10). The evidence for significant amounts of that formaldehyde state to be present is not clear-cut, however (10). Since small, but finite, nitrogen flow rate effects were observed (Figure 1) in the range employed in these experiments (0.5 NCM), some control of elution rate by gaseous formaldehyde diffusion through the shredded board or furnish particle-gas interface (vaporization) must have existed. Intraparticle diffusion limitations also seem likely at these particle sizes (~100 to 1,000 μm), although particle size effects were not observed in the high humidity weighing bottle tests with sizes below 180 μm. Intraparticle diffusion presumably involves methylene glycol, whose effective diffusion rate in the wood's water may well be decreased by strong interactions with cellulosics (perhaps reversible hemiformal reactions) during its passage to the particle surface.

Implications For Formaldehyde Emission From Large Panels. Much of the above discussion should be directly relevant to large panel emission. If intraparticle diffusion of methylene glycol is hindered under some conditions with comminuted materials, similar hindrance will exist in an actual board. Moreover, gaseous diffusion through particle-gas interfaces will be greatly slowed in a particleboard panel because no eluting gas is present to reduce the concentration

gradient and the interface layer thickness. In addition, the diffusion path to the panel surface will be tortuous, and panel surface-
air layer gaseous diffusion limitations may exist. Diffusion effects
are, therefore, undoubtedly very important in panel emission rates.

For a given board composition and structure the presence of diffusion limitations leads to lower emission rates and somewhat higher
internal concentrations of dissolved methylene glycol. That concentration increase may be sufficient to slow the net production of
formaldehyde via reversible hydrolyses, thereby lowering and pro-
longing the emission contributions from hydrolytic processes. Unfortunately, at the present state of knowledge we can only speculate
about which formaldehyde states in the board may be responsible for
emission at various points in the board's life. However, the water
extraction data (Figure 9) suggest the possibility of distinguishing
between "loosely held" formaldehyde (perhaps methylene glycol monomer
and oligomers and cellulose hemiformal) and more firmly bonded formaldehyde, the latter presumably including hydrolytic sources (perhaps UF, UF-wood, and cellulose formal). The shape of the UF board
curve in Figure 9 indicates that from 20 to 40 mg of formaldehyde per
100 g of board may belong in the "loosely held" category. In addition, the Perforator value for this board (11 mg/100 g) indicates
that it should meet the HUD and possibly the E-1 standards, and this
implies maximum emission rates at standard conditions between
2×10^{-5} and 9×10^{-5} mg per g board per hour (Appendix 3). Assuming
the "loosely held" formaldehyde (20 to 40 mg/100 g) is primarily
responsible for those emission rates, then leads to maximum times
required to dissipate those formaldehyde states, i.e., 3 to 6 months
at the HUD level and 1 to 2 years at the E-1 level. Continuing with
the argument, subsequently emitted formaldehyde should derive from
hydrolytic processes. Obviously, additional water extractions plus
measurements of actual emission rates on identical boards would be
needed to confirm this approach towards distinguishing formaldehyde
sources within boards.

Summary and Conclusions

This paper and a companion one (10) address the general question of
the source and mechanism of formaldehyde emission from bonded wood
products. I have restricted this paper to literature and original
FPL results derived from studies on wood-containing systems. The
companion paper, however, also includes literature and FPL results
related to (a) the chemistry and hydrolytic stability of formaldehyde resins and model compounds and (b) the reactions of formaldehyde and UF compounds with wood components and the hydrolytic stability of their products. For the sake of completeness I summarize
below the findings and conclusions from all three parts of the
companion paper.

Major Findings. The major findings are as follows:
        (a) In an acid-catalyzed UF-bonded board, formaldehyde can exist
in a wide variety of states. These states may include dissolved
methylene glycol monomer and oligomers, paraform, hexa, chemically
bonded UF resin states, chemically bonded UF-wood states (amidomethylene ethers with cellulose), cellulose hemiformals, and cellulose
formals.

(b) Each of those states is a potential source of formaldehyde emission by evaporation (methylene glycol) or by initial hydrolysis (all others). Unfortunately, we cannot now provide a complete listing of states in the order of their potential importance as emission sources. Clearly, however, some of the most weakly held states would be methylene glycol, cellulose hemiformal, amidomethylols, and cellulose amidomethylene ethers.

(c) In a base-catalyzed PF-bonded board, formaldehyde states may include: methylene glycol monomer and oligomers, chemically bonded PF resin states, chemically bonded PF-wood states, and cellulose hemiformals. Emission sources apparently include methylene glycol, cellulose hemiformals, and a PF resin state--possibly phenolic methylols.

(d) In Southern pine containing formaldehyde that was sorbed at room temperature and at the wood's natural pH or at pH 2 or 3, formaldehyde states may include methylene glycol monomer and oligomers and possibly cellulose hemiformals. These are all apparently readily removed from the comminuted wood at 80 percent RH (5 days) or in pH 3 water (0.2 day).

(e) Diffusion processes can very likely exert a major influence on emission rates from large panels. Depending upon board structure, composition, age, and exposure condition, emission-limiting diffusion steps may involve methylene glycol within the board's water or gaseous formaldehyde within the board or within the board-air interface.

Subsidiary Findings. The subsidiary findings are as follows:
  (a) Formaldehyde liberation from cured neat resins (PF and UF) is much greater than expected for those same resins cured in a particleboard, indicating that the wood alters the resin cure and/or the bondline pH or that diffusion effects predominate in the board.

(b) A cured PF resin liberates formaldehyde at significant rates that increase with humidity.

(c) The Perforator test measures formaldehyde in states that are present in cured neat PF and UF resins, in boards made with both resins, and in formaldehyde-sorbed wood. In all but the last, the Perforator values are much less than the amounts removable by simple exposure to high humidity.

(d) The limited durability of UF-bonded wood products probably results from the susceptibility of UF resin and UF-wood bonds to chain scission from both hydrolysis and swell/shrink stresses. In either case, formaldehyde is a likely product.

Acknowledgment

This work was partially funded by the Formaldehyde Institute. I am also greatly indebted to members of the Technical Committee of the Formaldehyde Institute for advice and for supplying materials. Professor James Koutsky, University of Wisconsin-Madison Chemical Engineering Department, and several of his students aided this effort with both advice and laboratory aid. Much of the FPL data were obtained by Ralph Schaeffer and Jill Wennesheimer.

Appendix 1.   Experimental Procedures

a. Formaldehyde Liberation By Weighing Bottle Technique. Ground,
sieved (-80 mesh or smaller) powder was weighed (10 to 150 mg) into a
glass weighing bottle (40 mm dia. x 40 mm high), a small glass cross
placed on the bottom of the container, a glass beaker (22 mm dia.
x 25 mm high) containing 5 ml of sulfuric acid or salt solution
placed on the cross, and the bottle sealed with its greased cap. The
assembly was then stored in a temperature chamber (usually at 27°C)
for a specified period at which time the beaker was removed and
replaced with a fresh solution. For a given weighed sample, the
solution was replaced no more than twice. At each removal the solution was analyzed for formaldehyde, usually by the chromotropic acid
procedure. Humidity in the sealed bottles was controlled by the
concentration of sulfuric acid or salt.

b.   pH. The ground sample (usually -80 mesh) was shaken with distilled water at a 1/10 ratio in a capped vial for at least overnight.
The pH of the supernatant was measured using a combination electrode.

c. Perforator Test. With unground particleboard the standard procedure (38) was followed in which about 100 g of 25 x 25 mm specimens
were refluxed in toluene for 2 hours with continuous extraction of
formaldehyde into water and subsequent analysis of the water for formaldehyde concentration. Analyses were by the acetylacetone fluorometric method (39). For ground resins and other materials, sample
amounts were adjusted to produce comparable formaldehyde concentrations.

d. Nitrogen Elution of Particleboard, Furnish, and Cured Resin.
25 x 25 x 16 mm specimens rested on a wire screen inside a horizontal glass tube (30 mm diam. x 750 mm long). Smaller particle size
material was placed either in a similar vertical tube, with bottom
gas feed, or in a continuously shaken Erlenmeyer flask, with gas feed
via a tube leading to the flask's bottom. Entering gas was preconditioned by passage through or over saturated salt solutions at room
temperature (23 ± 1°C). Exiting gas was continuously scrubbed of its
formaldehyde by passage through a series of impingers containing
water and held in ice water. The number of impingers in series
varied with gas flow rate and scrubbing time, based on prior experiments to establish conditions providing greater than 95 percent
scrubbing efficiency. At intervals the gas flow was interrupted to
allow changing to a series of fresh impinger solutions; the removed
impinger solutions were analyzed separately or after combination,
usually with the acetylacetone fluorometric method (39). A variety
of tests confirmed that no significant formaldehyde losses were
caused by adsorption on the polyethylene tubing or by leaks. A number of analyses by both the acetylacetone and chromotropic acid
methods showed no significant differences.
     Ground resin was eluted by nitrogen in a similar manner, the
primary exception being the use of only a few grams held in a glass
tube that contained sintered glass frits at both ends.

e. Water Extraction of Ground Wood or Board. Approximately 0.4 g of
ground (-80 mesh) sample were placed in a stoppered flask to which

were added 75 mL of water made to pH 3 with HCl and containing
100 mg/L sodium azide as bacterial inhibitor. The flasks were
shaken at 25°C and at intervals 10 mL aliquots were removed by
sucking through a sintered glass filter. At each removal, 10 mL
of fresh liquid were added to the flask through the filter; each
flask was sampled no more than three times. Aliquots were analyzed
by the fluorometric acetylacetone procedure (39).

Appendix 2. Materials

The UF particleboard was a commercial low emission product made
with a resin having an F/U ratio below 1.2. The PF board was an
experimental, industrial product, and the furnish was standard
industrial Southern pine material.
    Formaldehyde-sorbed Southern pine furnish was prepared by
allowing furnish to equilibrate for several days at room temperature
over water solutions of salts and formaldehyde, the salt serving to
control humidity. Formaldehyde-sorbed ground Southern pine was
similarly prepared except for a prior soaking with tartaric acid
solution (with sodium azide) at pH 2 or 3.

Appendix 3. Methods of Calculation

a. Rates for Particleboard Emission Standards. Assuming a steady
state condition for the concentration C_s in ppm of formaldehyde in
air and an emission rate ER from board in units of mg CH_2O per g dry
board per hour:

$$ER = K \frac{N}{L} C_s \qquad (A1)$$

where
    K = constant for conversion of units
    N = ventilation rate in hours
    L = board loading in m^2 exposed board area per m^3 of air space

	$C_s(25°C)$	N/L	ER
HUD (8)	0.3	1.2	9×10^{-5}
E-1 (7)	~0.12	1.0	2×10^{-5}

Literature Cited

1. Deppe, H-J. Holz-Zentralblatt, Stuttgart. 1982, 10, 123-124,
    126.
2. McVey, D. Proc. 16th Washington State Univ. Int. Symp. on
    Particleboard, Pullman, Wash., 1982.
3. Sundin, B. Proc. 16th Washington State Univ. Int. Symp. on
    Particleboard. Pullman, Wash., 1982.
4. Myers, G. E. Forest Prod. J. 1984, 34(5), 35-41.
5. Myers, G. E. Forest Prod. J. 1985, 35(6), 57-62.
6. Myers, G. E. Submitted to Forest Prod. J., 1985.
7. ETB Baurichtlinie. "Guideline for the Use of Particleboards
    with Respect to the Avoidance of Unacceptable Formaldehyde Con-
    centration in Room Air"; Commission for Uniform Technical
    Building Specifications, Beuth Publishers: Berlin, April 1980.

8. Department of Housing and Urban Development. Federal Register.
    1984, 49(155), 31996-32013.
9. Neusser, H.; Zentner, M. Holzforsch. Holzverwert. 1968, 20(5),
    101-112.
10. Myers, G. E. Proc. Symposium on Wood Adhesives in 1985: Status
    and Needs. Forest Prod. Lab. and Forest Prod. Res. Soc.
11. Hanetho, P. Proc. 12th Washington State Univ. International
    Symposium on Particleboard, 1978.
12. Christensen, R.; Robitschek, P.; Stone, J. Holz Roh-Werkst.
    1981, 39:231-234.
13. Marutsky, R.; Roffael, E.; Ranta, L. Holz Roh-Werkst. 1979,
    37, 303-307.
14. Myers, G. E. Forest Prod. J. 1984, 34(10), 59-68.
15. Kazakevics, A.A.R.; Spedding, D. J. Holzforschung 1979, 33,
    155-158.
16. Blomquist, R. F.; Olson, W. Z. Forest Prod. J. 1957, 7(8),
    266-272.
17. Neusser, H.; Schall, W. Holzforsch. Holzverwert. 1970,
    22(6), 116-120.
18. Robitschek, P.; Christensen, R. L. Forest Prod. J. 1976,
    26(12), 43-46.
19. Ginzel, W. Holz Roh-Werkst. 1973, 31, 18-24.
20. Higuchi, M.; Sakata, I. Mokuzai Gakkaishi 1979, 25(7), 496-502.
21. Dinwoodie, J. M. J. Inst. Wood Sci. 1978, 8(2), 59-68.
22. McNatt, J. D. Proc. Workshop on Durability of Structural
    Panels. Price, E. W., Ed., Pensacola, Fla., Oct. 1982,
    pp. 67-76.
23. Motoki, H.; Sagioka, T.; Tajika, K.; Sakai, T. J. Jap. Wood
    Res. Soc. 1984, 30(12), 995-1002.
24. Plath, E. Holz Roh-Werkst. 1973, 31(6), 230-236.
25. Sell, J. Holz Roh-Werkst. 1978, 36, 193-198.
26. Wilson, J. B. Proc. Workshop on Durability of Structural
    Panels, Price, E. W., Ed., Pensacola, Fla., 1982. pp. 53-57.
27. Ginzel, W. Holz Roh-Werkst. 1971, 8, 301-304.
28. Dinwoodie, J. M. Holzforschung 1977, 31(2), 50-55.
29. Marutzky, R.; Ranta, L. Holz Roh-Werkst. 1980, 38, 217-223.
30. Higuchi, M.; Shimokawa, H.; Sakata, I. Mokuzai Gakkaishi 1979,
    25(10), 630-635.
31. Kreibich, R. E.; Freeman, H. G. Forest Prod. J. 1970, 20(4),
    44-49.
32. Myers, G. E. Forest Prod. J. 1983, 33(5), 27-37.
33. Ezaki, K.; Higuchi, M.; Sakata, I. Mokuzai Kogyo 1982, 37(5),
    225-230.
34. Higuchi, M.; Kuwazuru, K.; Sakata, I. Mokuzai Gakkaishi 1980,
    26(5), 310-314.
35. Petersen, H.; Reuther, W.; Eisele, W.; WITTMAN, O. Holz
    Roh-Werkst. 1972, 30, 429-436.
36. Plath, L. Holz Roh-Werkst. 1967, 25(6), 231-238.
37. Casale, A.; Porter, R. S. "Polymer Stress Reactions,
    Vol. 1: Introduction," Academic Press: New York, 1978.
38. European Committee for Standardization. "Particleboards--
    Determination of Formaldehyde Content--Extraction Method Called
    Perforator Method"; Standard EN-120, CEN: Brussels, rev.
    March 1982.
39. Myers, G. E.; Nagaoka, M. Wood Sci. 1981, 13(3), 140-150.

RECEIVED January 14, 1986

Automated Flow Injection Analysis System for Formaldehyde Determination

Mat H. Ho

Department of Chemistry, University of Alabama, Birmingham, AL 35294

An automated and microprocessor-controlled flow injection analysis system was developed for formaldehyde emission measurements. This system was based on the modified pararosaniline method and a sampling rate of about 40 samples/hour was obtained. The relative standard deviations for sets of 15 repetitive measurements were 1.5% and 0.4% at concentrations of 1 and 10 μg/ml, respectively. The results obtained from this system correlated well with those obtained from the chromotropic acid. The simplicity, versatility, good precision, high sampling rate, and relatively low cost of the system make it attractive for the analysis of large numbers of formaldehyde samples.

Formaldehyde is a major component in the manufacturing of building materials such as particleboard, plywood and urea formaldehyde insulation. These materials can release formaldehyde vapor into the air of mobile homes, office buildings, and residences resulting in potential formaldehyde exposure to inhabitants and workers. It has been shown that formaldehyde in domestic air varies from near ambient concentrations (1-25 ppb) to as high as 4 ppm in new mobile homes (1). The health effects and possible carcinogenicity associated with formaldehyde exposure have created great concern on the monitoring of this chemical both in the workplace and indoor environments (2-5).

The monitoring and toxicological studies of formaldehyde exposure, as well as studies on the emission of this chemical from wood products generate large numbers of samples to be analyzed. Furthermore, it is necessary to monitor the emissions on a routine basis during production to ensure that the material continues to release low level of formaldehyde. In homes, particularly in mobile homes, the amount of formaldehyde release depends on the construction technoloy, ventilation, indoor temperature and relative humidity, and age, structure and porosity of building materials. It is, therefore, necessary to study the emision of formaldehyde from wood products as a function of these parameters.

0097-6156/86/0316-0107$06.00/0

The need for an automated and reliable system for formaldehyde determination is now clearly recognized. In response to this need, an automated and microprocessor-controlled flow injection analysis (FIA) system was developed in our laboratory. This system is based on the use of the modified pararosaniline colorimetric method (6). The simplicity, versatility, good precision, high sampling rate, complete automation and relatively low cost of the system make it attractive for the analysis of large numbers of formaldehyde samples. In this chapter, sufficient background in the principle of FIA will be presented to allow the readers to evaluate the technique and its potential application to the routine analysis of formaldehyde will be explored.

Principle of Flow Injection Analysis

Flow injection analysis (FIA), which was introduced by Ruzicka and Hansen (7-9) and by Stewart et al (10), is based on the concept of controlled dispersion of a sample zone when injected into a moving and nonsegmented carrier stream. In continuous flow analysis (CFA), successive samples are mixed and incubated with reagents on the way toward a flow through detector. The greatest difficulty to overcome in CFA was intermixing of adjacent samples during transport from the injection valve to the detector. In the past, it was widely believed that there are only two ways to prevent carryover in CFA: either by the use of turbulent flow or by air segmentation (11,12). Turbulent flow yields a flat velocity profile and therefore results in a lower sample zone dispersion than the laminar flow where the velocity profile is parabolic. However, it is difficult to obtain a turbulent flow in CFA. In the segmented CFA, air bubbles were used to divide the reaction stream into a number of compartments, thus preventing excessive dispersion of the sample by the dispersive sources inherent in the laminar flow (13). From this work the most popular automatic analyzer, the Technicon Auto-Analyzer, was developed.

Although the presence of air bubbles in the flowing stream creates several disadvantages, it was believed that air segmentation is essential for successful CFA. However, in 1975, Ruzicka and Hansen (7-9) and Stewart et al (10) demonstrated that continuous flow analysis can be performed in an unsegmented stream and the absence of the air bubbles actually offers several advantages. The name flow injection analysis (FIA) was proposed for this technique. A simple FIA system typically consists of a pump or some other means to propel the carrier and/or reagent, a sample injector, a reaction coil, a flow through detector and a recorder or data handling device. A precisely measured volume of sample is injected into a continuous flowing, nonsegmented carrier stream. The carrier stream transports the sample toward a flow through detector. Necessary reagents needed for a particular analysis are either present in the carrier stream or can be added further down stream on the way to the detector. As it moves towards the detector, the sample disperses into the carrier stream both longitudinally and radially by a combination of controlled laminar flow and molecular diffusion. The sample is mixed and reacted with reagents to form a detectable product which is then monitored by the detector. The response of the detector can be recorded in the form of sharp peaks as shown in Figure 1. These peaks reflect both the physical dispersion and chemical kinetics of the reaction that takes place between the injection port and detection point.

Dispersion is a phenomenon of great importance in FIA. When a liquid stream flows through a tube, the velocity of the liquid layer in contact with the tube's surface is practically zero and that at the center of the tube is twice the mean velocity of the liquid (12,14). From this stand point of the laminar flow, one can see that an injected sample bolus will result in a parabolic velocity profile (Figure 1). If a sample plug is placed into a moving stream, and if the longitudinal convection of the laminar flow is the only means of dispersion, it would have an infinitely long tail by the time it reached the detector. As a result, the carryover between adjacent injected samples becomes a serious problem in CFA. Fortunately, longitudinal convection is not the only means of dispersion. Molecules can diffuse, both longitudinally (in the direction of flow) and radially (perpendicular to the direction of flow), between the sample bolus and carrier stream. In the narrow tube and flowing stream, the contribution of longitudinal diffusion to the dispersion is less important than that of radial diffusion. Molecules at the walls of the tubes diffuse into the center of the sample zone. As a result, tailing of the sample due to parabolic velocity profile in the reaction tube is minimized by radial diffusion (Figure 1). Diffusion of molecules between the sample and carrier, the latter including reagent, explains not only the low carryover and high sample throughput but also the effective mixing of sample and reagents. Mixing between the sample and carrier due to dispersion is always incomplete, but because dispersion pattern for a given FIA system is perfectly reproducible, FIA yields precise results. The dispersion of the sample in the carrier stream is affected by several factors such as flow velocity, tube diameter, tube length and diffusion coefficient of the analyte. These parameters can be controlled in order to give an excellent reproducible dispersion. In FIA, dispersion is also frequently used to describe the degree of dilution of sample in the injector, reaction tube and detector. When sample is injected into the carrier stream, it travels as a gradually expanding plug which is slowly diluted by the carrier. Dispersion is required to provide adequate mixing of the sample and the reagent, however, increasing dispersion will decrease the analyte concentration and therefore reduces the sensitivity. Usually, dispersion is defined as a ratio of the concentration of the sample before mixing has occurred to the maximum concentration of the sample at the detector.

Since the reaction products are measured before steady-state conditions are established, the readout is available within seconds of introduction of the sample and FIA possesses the potential for high sample throughput. This technique has proven to be fast, precise, inexpensive, highly versatile and capable of automating a wide variety of wet chemical procedures. It is also possible to avoid or minimized the effect of interfering species in FIA because the reaction is not required to reach equilibrium. The tremendous interest in FIA in recent years is reflected by its substantial growth both in instrumental development and analytical applications

(11). There are several excellent reviews (12,15-17) and a book
(11) that describe the concept, principle, instrumentation,
applicability and limitation of FIA.

Experimental

Apparatus. Figure 2 shows the block diagram of the FIA system used
for the determination of formaldehyde. The system consists of a
sampler (Technicon, Tarrytown, NY), a peristaltic pump, a
microprocessor-controlled solution handling unit (Model SHS-200,
Fiatron Inc., Milwaukee, WI), a spectrophotometric detector (Model
LC 55, Perkin Elmer, Norwalk, CT) and a strip chart recorder. The
SHS-200 unit consists of a sample valve and a reagent valve
systems. The optical encoder, which is use for controlling the
pump speed, is mounted on the pump motor shaft to ensure precise
pump speed monitoring and regulation. The sample and reagent valve
systems consist of five three ways Teflon solenoid valves. All of
these are under software control and can be programmed via a front
panel keyboard (18). All parameters such as mode, pump speed,
washing time, sample injection time, time interval between
injections were programmed into the microprocessor control unit.
Several operational modes such as fixed sample volume, programmable
sample volume, programmable reagent volume, stop flow, merging
stream and on stream dilution can be obtained by programming the
pump speed, timing, and valve states (18). In this study, mode 20
was used and pararosaniline was allowed to flow continuously as
carrier stream. Formaldehyde samples were automatically fed into
the FIA system via a sampler which was also under microprocessor
control. The reaction coils consist of 650 cm of 0.8 mm i.d.
Teflon tubing and the temperature was controlled at 50°C by a
thermostated water bath. The flow rate was kept at 1.0 ml/minute.
This allowed about 196 seconds for the reaction to occur before
reaching the detector. The sample injection time was programmed in
order to inject 250 µl formaldehyde into the carrier stream.

Reagents. All chemicals were ACS analytical reagent grade and were
used without further purification. Deionized distilled water was
used for solution preparations. The stock pararosaniline reagent
was obtained as an 0.2% (W/V) solution in 1M HCl from CEA
Instruments, Emerson, NJ. The working pararosaniline solution (0.9
mM pararosaniline in 0.5 mM HCl) was prepared from the stock
solution and sufficient HCl was added to bring its concentration to
0.5 mM. The second reagent, which is 1.60 mM sodium sulfite, was
prepared by dissolving 0.2 g of anhydrous sodium sulfite (Fisher
Scientific Co., Fair Lawn, NJ) in deionized water and diluting to 1
liter. This reagent must be made fresh daily. Formaldehyde stock
solution, approximately 1 mg/ml, was prepared by diluting 2.7 ml of
37% formaldehyde solution (Fisher Scientific Co., Fair Lawn, NJ) to
1 liter with deionized water. The stock solution was standardized
using the sulfite method (19,20). This solution remained stable
for several months. Formaldehyde standard solutions were prepared
daily from the stock solution. A chromotropic acid solution, 0.01
g/ml, was prepared fresh by dissolving 4,5-dihydroxy-2,7-naphth-
alenedisulfonic acid disodium salt (Eastman Kodak, Rochester, NY)
in deionized water.

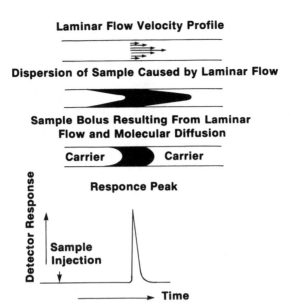

Figure 1. Dispersion of sample zone caused by laminar flow and molecular diffusion.

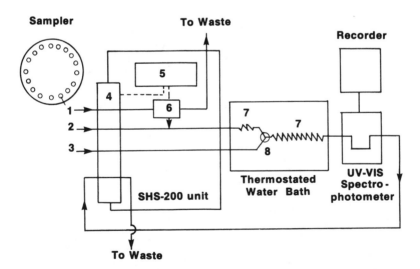

Figure 2. Schematic diagram of the microprocessor controlled FIA system for formaldehyde. (1) Formaldehyde standards or samples; (2) 0.9 mM pararosaniline in 0.5 M HCl; (3) 1.60 mM sodium sulfite; (4) peristaltic pump (5) microprocessor control unit; (6) sample injection valves system; (7) reaction coils; (8) Y connector

Procedure. Formaldehyde sample from the sampler was injected into the carrier stream where it was mixed with pararosaniline and then sulfite to form an alkylsulfonic acid chromophore which can be monitored spectrophotometrically at 570 nm. For calibration, standard formaldehydes were sequentially introduced after a stable baseline was obtained. At least five consecutively reproducible peaks were recorded for each concentration. After each study or each day of operation, the FIA system was cleaned to remove any pararosaniline film, alkylsulfonic acid colored product, or particulate matters. This reduced the scattered light in the absorption cell and the staining of the tubing walls. The clean-up procedure was initiated by running distilled deionized water through the system for five minutes followed by another five minutes washing with 0.1 N nitric acid and then flushing the unit for 30 minutes with deionized water. The chromotropic acid method was used for comparative studies, and the analytical procedure for the chromotropic acid method was based on the procedure recommended by the American Public Health Association (19).

Results and Discussion

The pararosaniline method has been used widely for the determination of formaldehyde in aqueous solutions and in the atmosphere. In this procedure mercury (II) - sulfite and acidified pararosaniline reagent were sequentially added to an aqueous formaldehyde solution (21,22). In 1965, an automated procedure for formaldehyde was described by Lyles et al (21). Later, Lahmann and Jander (22) modified the reagent concentrations to enhance sensitivity. This method has been adapted to the CEA 555 formaldehyde analyzer (CEA Instruments, Inc., Emerson, NJ). The major drawback of the pararosaniline method is the use of poisonous tetrachloromercurate to stabilize the sulfite reagent. In order to avoid the toxic hazard and disposal problem of mercury, a modified pararosaniline method for formaldehyde determination was developed by Miksch et al (6). To analyze a formaldehyde solution, the acidified pararosaniline reagent was added first and then sodium sulfite. Formaldehyde reacts with pararosaniline and sulfite to produce alkyl sulfonic acid which can be detected at 570 nm. Studies on the reagent stability, temperature dependence and interference of this method have also been published (23,24).

Concentrations of pararosaniline (0.9 mM), hydrochloric acid (0.5 mM) and sodium sulfite (1.60 mM) were selected to provide the same final concentrations after mixing as in the optimized conditions described by Miksch et al (6). No attempt was made to determine the pH of the reaction inside the flow system. Formaldehyde was injected into the stream of acidified pararosaniline and then merged with sodium sulfite to produce a colored product. The results were recorded as sharp peaks.

In the determination of formaldehyde using pararosaniline method, the temperature of the reaction should be controlled in order to obtain reproducible results (6,24). The rate of the reaction is also temperature dependent (6). In this study, the temperature of the reaction coil was kept constant at $50^{\circ}C$. Since Teflon is not a good thermally conductive material, it is expected that the temperature of the reaction was about $40^{\circ}C$.

Miskch et al (6) showed that the absorbance tends to decrease as the temperature increased above 25°C, probably because of the evaporation of sulfur dioxide from the acidic solutions. However, such sulfur dioxide or formaldehyde losses are not possible in our flow system due to containment of the sample and reagents within the Teflon tubing.

The sensitivity of the system, which was measured as peak heights, can be enhanced by increase the chemical development period following the addition of acidified pararosaniline and sodium sulfide. This can be done by increase the length of the reaction coil. The increase in residence time is counterbalanced, however, by an increase in the dispersion of the sample zone. The reaction coil of 650 cm was chosen for the FIA system. It is important to realize that only a relatively short residence time is achieved in FIA. Therefore, the FIA technique was originally not though to have a very wide scope of applications, since many colorimetric methods performed manually usually required 30 minutes or more for optimum color development. In the present case, optimum color development for formaldehyde determination using the modified pararosaniline procedure requires about 60 minutes at room temperature and 10-15 minutes at 40°C (6). In the FIA System, the chemical reaction never reached the steady state due to short residence time. However, the time is controlled precisely and excellent reproducible results can be obtained. Furthermore, mixing between formaldehyde and reagents due to dispersion may be incompleted, but because dispersion pattern for a given FIA system is perfectly reproducible, the system yields precise results.

Teflon tubing was used to construct the system. This reduced the staining of the tubing walls by pararosaniline and colored product. The staining process may increase the background or contribute to the memory effect following the analysis of high formaldehyde concentrations and therefore decrease the sampling frequency. Since the interferent studies has been reported elsewhere (6,25), it was not repeat here. However, it is expected that the selectivity in the FIA will be much better as compared to the manual procedure because FIA is a kinetic technique and the steady state is not allowed to achieve.

Figure 3 shows the typical response peaks of the FIA system for formaldehyde. The precision of all measurements was very good. The relative standard deviation for sets of 15 injections were 1.5% and 0.4% at concentrations of 1 and 10 µg/ml, respectively. Aqueous formaldehyde standards were used for the calibration. Linearity was observed for the concentration range from 1 to 15 µg/ml. The equation describing the linear portion of the calibration plot is given by $Y = 0.098 X + 0.031$ where Y is the peak height in absorbance unit and X is the concentration of formaldehyde in µg/ml. The calibration plot is shown in figure 4.

Comparison studies between the FIA and the chromotropic acid were performed. Fifteen samples with formaldehdye concentrations ranging from 1 to 10.8 µg/ml were determined by both methods and a correlation coefficient of 0.994 was obtained. This indicates a good correlation between two methods.

The flow injection system described here can be used for automated analysis of large numbers of formaldehyde samples. The

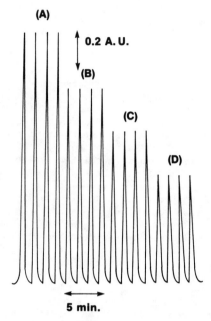

Figure 3. Typical response peaks of FIA system for formaldehyde.
(a) 13.5 μg/ml; (b) 10.5 μg/ml; (c) 8.0 μg/ml; (d) 5.0 μg/ml

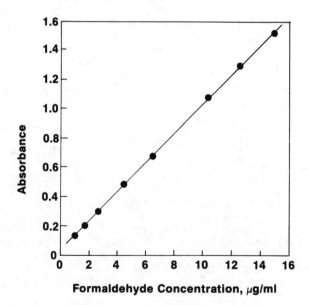

Figure 4. Calibration plot

sampling rate of the system is about 40 samples/hour. The sensitivity and detection limit of the system can be further improved by using the stop-flow (26) or pearl string reactor (27,28) techniques.

Acknowledgment

This work was supported by the University of Alabama at Birmingham Faculty Research Grant 2-12726.

Literature Cited

1. Consumer Product Safety Commission, Technical Workshop on Formaldehyde, April 9-11, 1980, National Bureau of Standards, Gaitherburg, MD.
2. "Formaldehyde: Evidence of Carcinogenicity", NIOSH Current Intelligence Bulletin No. 34, April 15, 1981.
3. Blackwell, M.; Kand, H.; Thomas, A.; Infante, P. Am. Ind. Hyg. Assoc. J. 1981, 42, A-34.
4. Bardana, E.J. Immun. Allergy Pract. 1980, 2, 11.
5. Gunby, P. J. Am. Med. Assoc. 1980, 243, 1697.
6. Miksch, R.R.; Anthon, D.W.; Fanning, L.Z.; Hollowell, C.D.; Revzan, K.; Glanville, J. Anal. Chem. 1981, 53, 2118.
7. Ruzicka, J.; Hansen, E.H. Anal. Chim. Acta 1975, 78, 17.
8. Ruzicka, J.; Hansen, E.H. Anal. Chim. Acta 1975, 75, 145.
9. Ruzicka, J.; Hansen, E.H. Anal. Chim. Acta 1978, 99, 37.
10. Stewart, K.K.; Beecher, G.R.; Hare, P.E. Anal. Biochem. 1976, 70, 167.
11. Ruzicka, J.; Hansen, E.H. "Flow Injection Analysis"; Wiley Interscience: New York, 1980.
12. Betteridge, D. Anal. Chem. 1978, 50, 832A.
13. Skeggs, L.T. Anal. Chem. 1966, 38, 31A.
14. Taylor, G. Proc. Roy. Soc. Ser. A 1953, 219, 186.
15. Rocks, B.; Riley, C. Clin. Chem. 1982, 28, 409.
16. Stewart, K.K. Anal. Chem. 1983, 55, 931A.
17. Ruzicka, J. Anal. Chem. 1983, 55, 1041A.
18. SHS-200 Operational Manual, Fiatron Systems, Inc., Milwaukee, WI.
19. "Methods of Air Sampling and Analysis", American Public Health Association: Washington, D.C., 1977, 2nd ed., pp. 303-307.
20. Walker, J.F., "Formaldehyde"; Reinhold Publishing Co.: New York, 1964; pp. 486-487.
21. Lyles, G.R.; Dowling, F.B.; Blanchard, V. J. Air. Pollut. Control. Assoc. 1965, 15, 106.
22. Lahmann, E.; Jander, K. Gesund-Ing. 1968, 89, 18.
23. Kuijpers, A.T.J.M.; Neele, J. Anal. Chem. 1983, 55, 390.
24. Georghiou, P.E.; Harlick, L. Anal. Chem. 1983, 55, 567.
25. Knox, S.E.; Que Hee, S.S. Am. Ind. Hyg. Assoc. J. 1984, 45, 325.
26. Ruzicka, R.; Hansen, E.H. "Flow Injection Analysis"; Wiley Interscience: New York, 1980; pp. 61-65.
27. Reijn, J.M.; Van der Linden, W.E.; Poppe, H. Anal. Chim. Acta 1981, 123, 229.
28. Reijn, J.M.; Poppe, H.; Van der Linden, W.E. Anal. Chem. 1984, 56, 943.

RECEIVED January 14, 1986

10

Enzymatic Methods for Determining Formaldehyde Release from Wood Products

Mat H. Ho and Jui-Lin Weng

Department of Chemistry, University of Alabama, Birmingham, AL 35294

Two sensitive fluorometric enzymatic methods for the determination of formaldehyde release from wood products were described. These methods were developed using the enzyme formaldehyde dehydrogenase to catalyze the oxidation of formaldehyde to form formic acid and NADH in the presence of oxidized nicotinamide adenine dinucleotide (NAD^+). The increase in NADH, which is directly proportional to the concentration of formaldehyde, is measured fluorometrically at λ_{ex}= 348 nm and λ_{em}= 467 nm. The NADH produced can also be reacted with resazurin in the presence of diaphorase to form resorufin, a highly fluorogenic compound. The fluorescence production is measured at λ_{ex} = 575 nm and λ_{em} = 590 nm. The optimal conditions as well as the sensitivity and linear range of these methods will also be described.

During the past decade, urea formaldehyde and phenol formaldehyde resin binders have contributed greatly to the progress of wood industries. Formaldehyde is widely used as a major component in the production of building materials, such as particleboard and plywood, and in urea formaldehyde foam insulation. However, the emissions of formaldehyde from these products create considerable concerns not only in the working environments but also in residences, mobile homes, and office buildings. These concerns have also been stimulated by reports on the health effects and carcinogenicity associated with formaldehyde exposure. Recently, numerous particleboard manufacturers have initiated programs to reduce formaldehyde release from their products, thus "low emission" urea formaldehyde resins were introduced (1,2). The emissions of formaldehyde from wood products have been addressed by several authors in this volume. This paper will focus on the development and application of two sensitive and specific analytical procedures for the determination of formaldehyde.
    The measurements of formaldehyde release from wood products usually involves two steps: sampling and analysis. For sampling,

0097–6156/86/0316–0116$06.00/0
© 1986 American Chemical Society

formaldehyde emissions were collected in water or sodium bisulfite absorbing solution using a suitable test such as large scale test chamber, mobile home simulator test chamber, quick test, or desiccator test (2). Chromotropic acid is the most widely used and recommended method for the analyzing of the collected formaldehyde. However, the chromotropic acid is potentially subjected to numerous interferences such as phenols, alcohols, olefins, aromatic hydrocarbons, nitrites, and nitrates (3,4).

Because of inherent interferences in the nonenzymatic reactions, such as chromotropic acid, there is a need for a more specific test which will yield a better estimation of actual formaldehyde levels release from wood products. The purpose of this paper is to introduce the use of an enzyme as an analytical reagent for formaldehyde determination and explore its potential utility for measuring formaldehyde emission levels. The use of an enzyme in the determination of formaldehyde is an attractive approach for a number of reasons including specificity and sensitivity. The tremendous progress in enzyme technology together with the advent of analytical instrumentation, encourages the use of enzymes for quantitation of various substrates, inhibitors, activators and enzymes themselves. The growing analytical applications of enzymes has been reflected in extensive publications in recent years (5,6), with most of these applications in clinical chemistry. Enzymes have found little or no practical application in environmental chemistry. This work represents the first attempt to use enzyme for the specific and sensitive determination of formaldehyde.

Principle of Enzymatic Method for Formaldehyde Determination

Enzymes are proteins which have the capability to catalyze many complex chemical reactions. Outstanding properties of these biological catalysts are their specificity and their capability of catalyzing the reaction of a substrate at very low concentration. Many enzymes are specific for a particular reaction of a particular substrate even in the presence of other isomers of that substrate or similar compounds. Some other enzymes are specific for a particular class of compounds.

In 1974, Uotila and Koivusalo (7) reported that the oxidation of formaldehyde to formate can occur in all tissues, and formaldehyde derived from methanol appears to be oxidized by glutathione-dependent formaldehyde dehydrogenase in the cytosol. Cinti et al. (8) showed that formaldehyde derived from the microsomal N-demethylation reactions is oxidized by a non-glutathione-requiring formaldehyde dehydrogenase in the mitochondria. In this study, a non-glutathione-dependent enzyme was used.

Two novel fluorometric methods for the determination of formaldehyde were developed using the enzyme formaldehyde dehydrogenase. The principle of these methods is based on the quantitative oxidation of formaldehyde with nicotinamide adenine dinucleotide (NAD$^+$), catalyzed by formaldehyde dehydrogenase, to form formic acid and NADH as shown in the following reaction:

$$\text{Formaldehyde} + \text{NAD}^+ \xrightarrow{\text{Formaldehyde hydrogenase}} \text{Formic acid} + \text{NADH} \quad (1)$$

In the fluorometric method I, the NADH produced is monitored spectrofluorometrically at an excitation wavelength (λ_{ex}) of 348 nm and an emission wavelength (λ_{em}) of 467 nm. The fluorescence intensity is proportional to the concentration of formaldehyde. Alternatively, the following coupled reaction can be used for more sensitive analysis of formaldehyde in the ppb concentrations:

$$\text{NADH} + \text{Resazurin} \xrightarrow{\text{Diaphorase}} \text{NAD}^+ + \text{Resorufin} \quad (2)$$

The NADH produced in reaction 1 is in turn oxidized by resazurin. This reaction is catalyzed by diaphorase which acts as an electron carrier. The reduced form of resazurin is a highly fluorogenic compound called resorufin. The fluoresence production is measured at λ_{ex} of 575 nm, and λ_{em} of 590 nm, and is linearly proportional to the concentration of the formaldehyde.

The concentrations of formaldehyde participating in these enzymatic reactions can be determined by two different methods: the equilibrium method and the kinetic method (5,6). In the equilibrium method, the reaction is allowed to go to completion and the product formed is measured, provided the product is chemically and/or physically distinguishable from the substrate. NADH in the enzymatic method I and resorufin in the enzymatic method II are measured fluorometrically and they are proportional to the concentration of formaldehyde. The equilibrium method is generally more precise and reliable, particularly in the manual and non-automated procedures. However, this method requires a large amount of enzyme to ensure relatively rapid reaction; otherwise the time required to reach equilibrium becomes relatively long. In the kinetic method, the initial rate of the enzymatic reaction is measured without waiting for the reaction to go to completion. The initial rate method is fast, however, temperature, pH and ionic strength of buffer, stirring rate and timing must be carefully controlled for good results. If the time required to reach equilibrium is long, large quantity of enzyme is needed and in this case the kinetic method is preferred over the equilibrium method.

Method and Procedure

Reagents. Formaldehyde dehydrogenase solution, 10 units/ml, was prepared in phosphate buffer (pH 7.5, 0.1M). Formaldehyde dehydrogenase (EC 1.2.1.1) from Pseudomonas putida was obtained from Sigma Chemical Co., St. Louis, Missouri. Oxidized nicotinamide adenine dinucleotide (NAD$^+$) solution, 5 mg/ml, was prepared using doubly distilled deionized water. Diaphorase Solution, 72 units/ml, was prepared in phosphate buffer (pH 7.5, 0.1M). NAD$^+$ and diaphorase (EC 1.6.4.3, from Clostridium kluyveri) were also obtained from Sigma. Formaldehyde dehydrogenase, NAD$^+$, and diaphorase solutions should be prepared fresh daily and stored at 4°C when they are not in use. Resazurin was dissolved in doubly distilled deionized water to give a final concentration of 30 mg/l solution in a dark bottle.

Resazurin was purchased from Aldrich Chemical Co., Milwaukee, Wisconsin. Formaldehyde stock solution was prepared by diluting 2.7 ml of 37% formaldehyde solution to 1 liter with deionized water and standardized using the sulfite method (3,9). This solution remained stable for several months. Formaldehyde solution was ACS reagent grade and obtained from Fisher Scientific, Pittsburgh, Pennsylvania. Formaldehyde standard solutions for the calibration were prepared daily from the stock solution. Other chemicals for formaldehyde standardization and buffer preparations were all analytical reagents and were used without further purification.

Apparatus. Fluorescent measurements were made with an AMINCO SPF-125 spectrofluorometer (American Instrument Co., Silver Spring, Maryland) equipped with a thermostated cuvette. A strip chart recorder (Omnigraphic-2000, Houston Instrument, Austin, Texas) was used to record the fluorescent intensity as a function of time. Temperature was controlled with a LAUDA thermostated water bath circulator (Model K-2/R, Fisher Scientific Company, Pittsburgh, Pennsylvania).

Analytical Procedure. For the enzymatic method I, 0.98 ml of phosphate buffer (pH 8) and 50 μl of formaldehyde dehydrogenase were pipetted into a cuvette. To this 400 μl of formaldehyde sample, or standard, were added, and mixed by shaking for 5 seconds. The cuvette was placed in the spectrofluorometer (λ_{ex}= 348 nm and λ_{em} = 467 nm) and a stable baseline was obtained before proceeding. The reaction was initiated by injecting a 50 μl solution of NAD$^+$ into the cuvette with the increase in fluorescence recorded as a function of time. The fluorescent intensity was measured one minute after injection, or at the steady state.

For formaldehyde analysis using method II, 0.83 ml of phosphate buffer was pipetted into a sample cuvette. To this 50 μl of formaldehyde dehydrogenase, 50 μl of diaphorase, and 100 μl of resazurin were added. Next 400 μl of formaldehyde sample, or standard, were added, then mixed by shaking for 5 seconds. The cuvette was placed in the spectrofluorometer (λ_{ex} = 575 nm and λ_{em} = 590 nm) and a stable baseline was obtained before proceeding. The reaction was initiated by the addition of 50 μl of NAD$^+$ solution to the cuvette, with the fluorescence intensity measured one minute after injection, or at the steady state. The increase in fluorescence was also recorded as a function of time.

Results and Discussion

Enzymatic Fluorometric Method I. There are several factors, such as enzyme concentration, substrate concentration, pH of buffer, and temperature, which can affect the kinetics of the enzyme catalyzed reaction. These factors should be optimized and carefully controlled in order to obtain the most sensitive and reproducible results. The results of the optimization studies are summarized in Table I.

Figure 1 shows the plots of the fluorescence intensity versus time for several different concentrations of formaldehyde. About 75% of the fluorescence can be obtained within the first minute and

Table I. Optimal Concentrations of the Reagents Used in the
          Enzymatic Fluorometric Methods

| | Amount/Determination | |
Reagent	Method I	Method II
Formaldehyde Dehydrogenase	0.50 units	0.50 units
NAD$^+$	0.25 mg	0.25 mg
Diaphorase	--	3.6 units
Resazurin	--	3.0 μg

the steady state is achieved in about 5 minutes.

After the optimal conditions of the assay were investigated, a series of calibration plots were prepared using different formaldehyde concentrations. Figure 2 shows typical calibration plots for 30 seconds, 1 minute, and at the steady state. If the fluorescence was measured at the steady state, the calibration curve fits the general equation Y = 69.67(X) + 9.82 where Y is the fluorescent intensity and X is the corresponding formaldehyde concentration. A linear dynamic range was observed up to 0.54 μg/ml. The lowest concentration of formaldehyde in the assay solution which can be determined with this method is limited by experimental reproducibility and instrumental resolution, which was found to be 0.02 μg/ml. Figure 2 also shows the calibration plots in which fluorescence was measured at 30 seconds and 1 minute after the reaction had started, and the data fit the following equations: Y = 60.87 (X) + 7.58 and Y = 43.18 (X) + 5.38, respectively.

Enzymatic Fluorometric Method II. In this method, the concentration of formaldehyde dehydrogenase, diaphorase, NAD$^+$, resazurin and the pH of buffer were optimized. The results of the optimized parameters are also shown in Table I. The times required to obtain the steady state (of about 3 minutes) at different formaldehyde concentrations are shown in Figure 3.

The calibration curve was obtained using optimized concentrations of formaldehyde dehydrogenase, NAD$^+$, diaphorase, resazurin and buffer pH. The calibration curve measured at 1 minute after injection fits the equation Y = 120 (X) + 4.68 as shown in Figure 4. This figure also shows the extended calibration plot at low concentrations and the data fit the equation Y = 0.437 (X) + 11.3. The lowest concentration of formaldehyde in an assay solution which can be determined with this method is 0.27 ng/ml.

The slopes of the calibration plots, 60.87 fluorescence unit per μg/ml for enzymatic fluorometric method I and 120 fluorescence unit per μg/ml for enzymatic fluorometric method II, show that method II is approximately twice as sensitive as method I. This is due to formation of the intensely fluorogenic resorufin in method II. The higher sensitivity and lower detection limit of the enzymatic fluorometric method II will have potential applications in air sampling of formaldehyde emissions since sampling time can be reduced.

Several inorganic and organic compounds such as nitrite, nitrate, phenols, alcohols, organic solvents, and aromatic hydrocarbons are known to be interferents in the chromotropic acid method were investigated. No interferences were observed from these compounds even at high concentration (1,000 μg/ml). Some

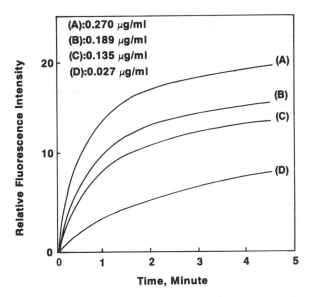

Figure 1. Plots of fluorescence intensity versus time in enzymatic method I.

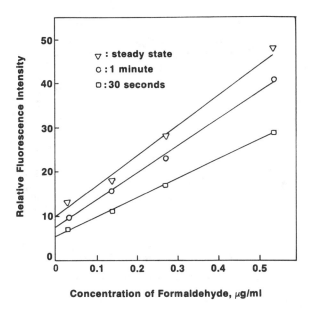

Figure 2. Calibration curve for the enzymatic fluorometric method I.

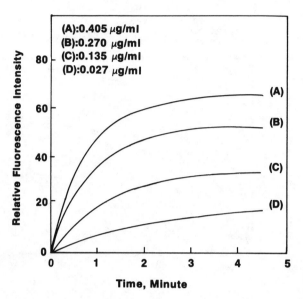

Figure 3. Plots of fluorescence intensity versus time in enzymatic method II.

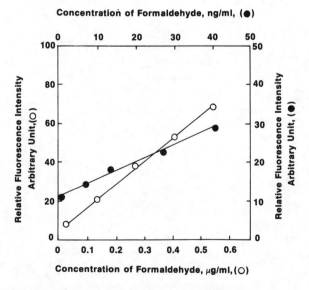

Figure 4. Calibration plots for the enzymatic fluorometric method II in ppm and ppb levels.

higher   aldehydes,   such   as   acetaldehyde,   propionaldehyde,
crotonaldehyde,   benzaldehyde   and   acrolein,   slightly   interfere   at
high concentrations.

Use of Enzymatic Methods for Determination of Formaldehyde Emission
from Wood Products.

The   measurement of formaldehyde release from wood products involves
the   collection   of   formaldehyde   vapor   in   the   test   chamber   using   a
suitable   absorbing   solution   and   then   analyzing   the   formaldehyde
collected.   For many years, formaldehyde emission measurements were
carried   out using the desiccator test sampling method (1,10) due to
its   simplicity.   In   this   method,   specimens   of   particleboard   or
paneling,   after   conditioned overnight at 25°C and 50% relative
humidity,   are   placed   in   a   clean,   dry   desiccator   containing
distilled   water.   For 24 hours test, 300 ml of distilled water were
used   in place of 25 ml used in the 2 hours test.   At the end of the
testing   period,   the   water   solution   is   analyzed   for   formaldehyde
content.   Recently, Lehmann (2) investigated many test procedures
such   as large scale test chamber, mobile home simulator test, quick
test,   quick   air test and desiccator test, and found that the large
scale   test   chamber   is   the   most   accurate   and   reliable   means   of
estimating formaldehyde emission from wood products.
      These   test   chambers   can   be   incorporated   to   the   enzymatic
methods   for   formaldehyde   determination.   Formaldehyde   emissions   of
a   product,   or mix of products, to the ambient air can be collected
in   distilled   water   or   1%   sodium   bisulfite   as   the   absorbing
solution.   After   collection,   formaldehyde   samples   are   analyzed   as
described   above.   In   the   mobile   home   simulator   test   method   (2),
double   or   triple   impingers, which are placed in series, should be
used   in   order   to collect all of the formaldehyde vapor.   The test
conditions   should simulate the actual environment.   Several factors
such   as   temperature   and relative humidity of the system including
the   specimens   and   background of formaldehyde in the test chamber,
affect   the   precision   and   accuracy   of   the   results.   It   has   been
shown   that   a   7°C   change   in   temperature   doubles   the   emission
level   (1).   The   temperature   of   the   test   chamber   should   be
maintained   at ± 0.1°C.   Since formaldehyde in aqueous solutions
is   unstable,   all   samples should be analyzed within one hour after
collection.
      The   enzymatic   methods   described   in   this   paper   are   not   only
more   specific   but   also   more sensitive than the chromotropic acid
method.   These   methods   can   be   used   for   the   measurement   of
formaldehyde   emission   from   wood   products   as well as formaldehyde
exposure in the workplace and in indoor environments.

Conclusion

We   have   developed two novel new enzymatic fluorometric methods for
the   trace   analysis   of   formaldehyde.   Due   to   their   simplicity,
sensitivity   and   specificity,   these   methods   should   find   wide
applications   in   the   monitoring of formaldehyde released from wood
products.   As we stated above, enzymatic fluorometric method II
does   offer   higher   sensitivity   and   better   detection   limit   over
enzymatic   fluorometric   method   I.   However,   method   II   requires   two

enzymes and is more expensive than method I, which only uses one enzyme. So the choice between use of method I or method II depends upon your need. If you are not concerned about the sensitivity and the low detection limit, you may simply use method I. Furthermore, the enzymes can be immobilized and can then be reused many times, up to several hundred assays, thus substantially reducing the cost of analysis. An obvious application of the immobilized formaldehyde dehydrogenase is in the automated flow injection system for analysis of large numbers of environmental samples. Such extension of the work described here is already in progress in our laboratory.

Acknowledgments

This work was supported by the University of Alabama at Birmingham, Faculty Research Grant 2-12726.

Literature Cited

1. McVey, D. T. Proc. 16th Wash. State Univ. Int. Sym. Particleboard, 1982, p. 21.
2. Lehmann, W. F. Proc. 16th Wash. State Univ. Int. Sym. Particleboard, 1982, p. 35.
3. "Methods of Air Sampling and Analysis," American Public Health Association: Washington,D.C.,1977, 2nd ed.,pp.303-307.
4. Krug, E. L. R.; Hirt, W. E. Anal. Chem. 1977, 98, 1865.
5. Guilbault, G. G. "Handbook of Enzymatic Methods of Analysis"; Marcel Dekker: New York, 1976
6. Carr, P. W.; Bowers, L. D. "Immobilized Enzymes In Analytical and Clinical Chemistry"; John Wiley & Sons: New York, 1980.
7. Uotila, L. Koivusalo, M. J. Biol. Chem. 1974,249, 7653.
8. Cinti, D. L.; Keyes, S. R.; Lemelin, M. A.; Denk, H.; Schenkman, J. B. J. Biol. Chem. 1976, 251, 1571.
9. Walker, J. F. "Formaldehyde"; Reinhold Pub. Co.: New York, 1964; p. 486-7.
10. Newton, L. R. Proc. 16th Wash. State. Univ. Int. Sym. Particleboard, 1982, p. 45.

RECEIVED January 14, 1986

A Model for Formaldehyde Release from Particleboard

J. J. Hoetjer and F. Koerts

Methanol Chemie Nederland VoF, postbus 109, 9930 AC Delfzijl, The Netherlands

In cooperation with DSM, MCN developed a method of measurement for the determination of the formaldehyde release from particle board, based on a theorie for mass transfer, implying that under steady state conditions the emission of formaldehyde of a given particle board can and should be defined by two parameters of the particular board. These two parameters are (1) C_e; defined as the equilibrium formaldehyde concentration (with ventilation rate "0") and (2) k_{og}; defined as the overall mass transfer coefficient of the board. In (ideal mixed) climate rooms the stationary formaldehyde concentration (C_g) as function of the ventilation rate (n) and load factor (a) is given in the relation:

$$1/C_g = 1/C_e + n/(C_e . k_{og} . a)$$

Plotting $1/C_g$ against n/a, gives a straight line, from which both concerned board properties are gathered. Graphs show that independent of the size of the apparatus, this statement is backed up quite well. Various examples that influence both those parameters illustrate the use of this formaldehyde emission method.

In various countries requirements and rules for the release of formaldehyde by particle board are being specified. On drawing up these rules, it is often desirable that they be related to a maximum admissible concentration in living environments.

For that purpose various institutes have made attempts to develop tests to characterize the release of a given particle board. These methods all have in common that they represent the emission with one and only one characteristic value.

First of all, the aim of this lecture is to demonstrate that it is possible to describe the formaldehyde emission in an acceptable manner with two characteristic particle board parameters, whilst this is not possible on the basis of only one characteristic and therefore neither on the basis of a test giving only one value.

0097–6156/86/0316–0125$06.00/0
© 1986 American Chemical Society

The second issue concerns the relation between climate chambers
and living environments. We wish to make clear that it is not
absolutely necessary to determine the two parameters using large
climate chambers as have been installed here and there and also
that with good provision for air circulation they can indeed be
regarded as well defined systems, suitable for the determination
of the particle board in question.

However, such environments must not be regarded as ideal or
standard living environments. In practice, living environments
present us with conditions that are much less well defined and may
vary among themselves, which, by definition, make them unsuitable
for the determination of the above mentioned particle board
parameters.

On the other hand, when once the two particle board
parameters have been measured in a suitable way, it is
fundamentally possible to calculate the expected formaldehyde
concentration, that is, at the same temperature and relative
humidity. Even then an estimation for living environments can be
made.

The third issue concerns combinations of different boards.
Later an example will show how the formaldehyde concentration can
actually be calculated for an environment with several emission
sources.

The development of the various mathematical equations are
given in the enclosures. Details concerning the apparatus and the
way in which it was used in the determinations can be found
elsewhere. (3)

For the purpose of our study it is assumed that the
temperature and the relative humidity are constant. In the
practical examples these values have throughout been kept at 20°C
65% relative humdity.

Introduction to the model

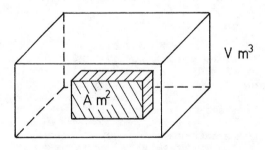

$$a = A/V = \text{loading factor} \quad m^2/m^3$$
$$C_g = CH_2O \text{ concentration} \quad mg/m^3$$

Figure 1. Particleboard in an enclosed space.

When we place a piece of particleboard with a surface area of A m^2 in an enclosed space with a volume of V m^3, in which at time zero no formaldehyde gas is present in the air (Figure 1), it is known that the particle board will release formaldehyde into the air and that, viewed over a period of time, the rate of release will not be constant but decreases as the formaldehyde concentration C_g in the environment increases, until a certain maximum concentration has been reached. (Figure 2)

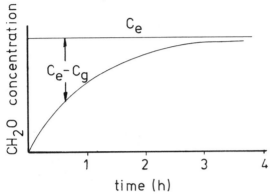

Figure 2. Formaldehyde concentration as a function of time without ventilation.

Something similar occurs with the vapour pressure of water, when a tray of water is placed in a dry enclosed space. After a period of time the vapour pressure will reach its maximum, a 100% relative humidity. Obviously the time required for this maximum vapour pressure to be reached, depends on three factors:
- the area of the surface;
- the extent to which the air is interchanged to equalise the (formaldehyde) concentration in the bulk of the gas phase;
- the nature of the interface.
    In the case of the tray of water for example, this last factor might be visualised as affected by impurities at the water surface. This surface might even be entirely covered up by paraffin, analogously to the behaviour of a painted particle board. The surface has, so to say, a certain resistance for mass transfer. The reciprocal value of this resistance is called "mass transfer coefficient".
    Returning to our tray of water, part of the resistance is on the side of the liquid phase. This is the resistance which has to be overcome by the molecules, to get to the surface of the water and penetrate the surface.
    Another part is on the side of the gas phase, namely the resistance which has to be overcome by the molecules to get from the interface into the bulk of the gas phase. For the time being the considerations will be restricted to the overall mass transfer of the particle board concerning the gas phase, here called k_{og}.

It must be made sure that at the experimental set up and in
carrying out the measurements, the gas side resistance is
neglectable, so that the mass transfer coefficient to be measured,
can be entirely contributed to the particle board. This aspect
will be discussed again, when the difference between a climate
chamber and a living environment will be discussed.

The curve shown in Figure 2 is a logarithmic function, which
means that a straight line is obtained, if the logarithm of the
driving force - this is the difference between the equilibrium
concentration and the current formaldehyde concentration ($C_e - C_g$)
- is plotted against time.

From the point of intersection of this straight line with the
Y-axis, together with the slope of this line, C_e and k_{og} can be
calculated, as:

$$Ln (C_e - C_g) = Ln C_e - k_{og} . a . t$$

in which $a = A/V$ m^2/m^3
and $1/k_{og}$ = mass transfer resistance (m/s).

There are more possibilities for measuring the two parameters
and in principle there are two models to calculate the
formaldehyde emission parameters. The two models that can be
applied are the "ideal mixing" model (Figure 6) and the "plug
flow" model (Figure 3).

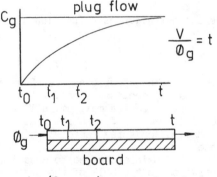

$$Ln (C_e - C_g) = Ln C_e - k_{og}.a.V/\emptyset_g$$

with $\emptyset_g$ = rate of airflow (m^3/s).

Figure 3. Plug Flow model.

Plug flow model

If air, containing no formaldehyde, is passed over a channel,
which is placed above a particle board surface, the relation
illustrated in Figure 3 is obtained. The air passing over the
particle board becomes increasingly rich in formaldehyde. If only
the channel is long enough the equilibrium concentration will be
reached again. The air flowing over the channel remains in contact
with the particle board for a given period, the residence time.

Increasing the rate of airflow means decreasing the residence time and vice versa. Measurements of the formaldehyde concentration in the exit air will thus give information on the driving force as a function of residence time. The mathematical equations underlying the calculation are shown in Figure 3.

Determination of the equilibrium concentration

It is important that the measurements always be completed with a measurement of the equilibrium concentration as such. This can be done by using a gasburette, like this is pointed out in Figure 4

ventilationrate n = 0

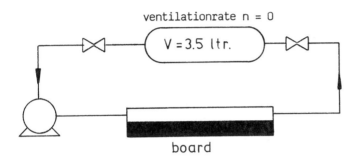

board

Figure 4. Determination of equilibrium concentration.

Every particleboard has its own characteristic maximum formaldehyde equilibrium pressure (C_e). This equilibrium concentration moreover depends on the temperature and the relative humidity.

Ideal mixing model

One has to make sure that there is sufficient turbulence or mixing in the experimental set up. Otherwise the principles at the basis cannot be applied. For example, the height of the channel should not be too great, unless provisions be made to achieve another well defined measuring system. Figure 5 illustrates this point.

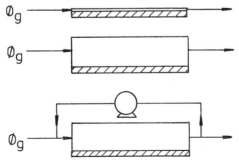

Figure 5. Different flow situations.

Without special provisions, an undefined flow model would be obtained, as a matter of fact much similar to a normal living environment encountered in practice.

Such a broad channel as shown in the middle of Figure 5, can however easily be changed back into another, again well defined flow model, by thouroughly mixing the air in it. If the room is too small to contain an electric fan, the air can also be mixed by using an externally applied circulation. This circulation should be a multiple of the gasflow. See the lowest example on Figure 5.

This model can be indicated as the so called "ideal mixing" model, as is given schematically in Figure 6.

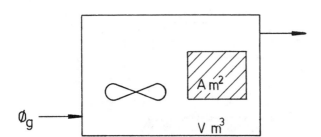

$$1/C_g = 1/C_e + \emptyset/k_{og}.A.C_e = 1/C_e + n/k_{og}.a.C_e$$

$$n = \emptyset_g/V \quad s^{-1} \quad \text{ventilation rate}$$

Figure 6. Ideal mixing model.

The formulae given here can be derived from the mass balance (see enclosure 2). For this model the reciprocal values of the formaldehyde concentration are plotted along the Y-axis of a graph and the corresponding airflow, eventually devied by the volume - the so called ventilation rate - is plotted along the X-axis. Again a straight line is obtained, from which both the parameters can be derived. It is inferable from the formulae that the volume of the test chamber is not essential. This too will be illustrated later (see Figure 7).

Illustrations

How things can go wrong, when the system is not sufficiently defined, is illustrated in Table I.
Three situations are shown. In each of them the concentration of formaldehyde in the exit air has been measured for four rates of airflow. The equilibrium value of the examined particleboard sample has been determined as well (1.06 mg/m^3).
Situation 1 : In this case the channel was 5 mm high.

Table I. Formaldehyde Concentration as a Function of Airflow
for Three Situations

airflow
$\phi_g \times 10^5$

	1	2	3

m^3/s CH_2O conc. mg/m^3

0	1.06	1.06	1.06
2.62	0.21	0.13	0.23
1.68	0.30	0.17	0.29
1.23	0.38	0.22	0.40
0.32	0.74	0.27	0.70

Situation 2 : Here the same rates of flow were used for a channel
50 mm high, resulting in much lower airspeeds.
Situation 3 : Measurements were taken at the channel with the same
height as in situation 2, yet with applying external
circulation, which implies that much greater air
velocities have been realized again.
From the results it can be concluded that the formaldehyde
concentrations in the exit air in situation 2 differ from those in
situations 1 and 3, which are almost the same. The reason is that
in situation 2 the exchange of formaldehyde between the
particleboard and the air concerned, was not complete. So the
measurements in situation 2 do not fit in with the equilibrium
determined.
      With application of the two mentioned mathematical models,
the two formaldehyde parameters for the three situations can be
calculated. The results are given in Table II.

Table II. The Calculated Board Parameters

	1		2	3	
	plug flow	ideal mix.		plug flow	ideal mix.
C_e	0.95	1.11	0.35	0.90	1.01
$k_{og} \times 10^4$	5.6	4.3		4.9	4.8
r	0.99	0.99		0.99	0.99

the measured $C_e = 1.06$ mg/m^3.

The calculated values of the mass transfer parameters for both the
flow models with the results of the situations 1 and 3 are shown.

The two models produce slightly differing results. In general,
application, of the ideal mixing model gives the most satisfactory
results, independant of the actual flow situation. There is indeed
a theoretical explanation for this. Therefore the model for ideal
mixing is usually applied. It is always necessary to check whether
the straight obtained, does in fact pass through the measured
equilibrium point or at least comes close to it. In situation 2
this is clearly not the case. Conclusion : Situation 2 does not
fit in with the model.

Out of the results of the intersection should follow an
equilibrium concentration of 0.35 mg/m^3, which is not in
accordance with the determined equilibrium value. So this
experimental set up is a case of a situation which is not well
defined and therefore not suitable for measurement of the relevant
formaldehyde release parameters of the particleboard.

To explain this, it can be argued that a not inconsiderable
increase in resistance to mass transfer has been set up in the gas
phase, which in fact may vary from situation to situation. Such
situations are indeed normal in everyday practice. This explains
why in practice, especially at low ventilation rates, much lower
concentrations are found, than would follow from measurements done
in climate chambers with good circulation. Such intensive
circulations remain absolutely necessary if determination of the
characteristic particleboard parameters is wanted, independant of
the test environment.

The formaldehyde concentration measured in situation 2 (see Table
I) can easily be explained by introducing an extra mass transfer
resistance for the air, which by the way, depends on the
ventilation rate as well. The extra mass transfer resistance of
the air decreases with increasing ventilation rate. The reason for
this is that the ventilation rate also influences the circulation.
The extra mass transfer resistance can be expressed by the
formulae:

$$1/k_p = 1/k_{og} + 1/k_{air}$$

in which $1/k_p$ = resistance in practice
and $1/k_{og}$ = resistance of the board
and $1/k_{air}$ = resistance of the air (in living environments)

Quantitative values of the mass transfer resistances

For bare particleboards in suitable test chambers, mass transfer
resistances are usually found to lie between 1,500 and 10,000 s/m.
When there is no internal circulation or when there is
insufficient turbulence, it is not uncommon to find an extra mass
transfer resistance for the gas phase of 12,000 s/m at a
ventilation rate of 0.75 per hour. A more detailed estimation is
given in the summary.

Independence of the volume

An other statement that should be illustrated, is the fact that neither the volume of the test chamber nor the loading factor influences the results found for the two parameters.

In figure 7 the ideal mixing model is applied for two different test chambers. Climate chamber A had a volume of 52 m^3 and a loading factor of 1 m^2/m^3. Climate chamber B had a volume of only 75ml and a loading factor of 200 m^2/m^3. It can be seen that the results obtained are in good agreement.

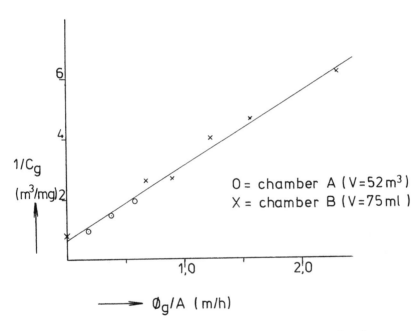

Figure 7.   Results are independent of loading and volume.

Examples

Example 1 : molar ratio

One of the important parameters in producing urea formaldehyde resins with a low formaldehyde level, is the so called molar ratio. Table III shows that the parameter C_e is closely related to the molar ratio, which varies from 0.70 to 1.30. The mass transfer coefficient is not related to the molar ratio, while this parameter in principle is only related to the nature of the surface.

Table III. Relation of Molar Ratio and Equilibrium Concentration

molar ratio F/U	C_e mg/m^3	k_{og} x 10^4 m/s
0.7	0.11	1.4
0.9	0.27	1.8
1.0	0.24	1.7
1.15	0.34	1.6
1.3	0.65	2.0

Remark : The mechanical properties of the concerned particleboards are not comparable.

Example 2 : aging

Table IV. Equilibrium Concentration in Course of Time

age days	C_e mg/m^3	k_{og} x 10^4 m/s	slope m^2.s/mg
X + 1	1.8	1.2	4500
2	1.7	1.3	4670
3	1.5	1.5	4520
7	1.3	1.2	6400
8	1.0	1.5	6800

In Table IV the results of a sample investigated a few days after the production of the board are given.

It is sometimes thought that it is the slope as such to be a board characteristic. But yet it can clearly be seen here that, in spite of the slope varying, the difference in the release of formaldehyde in the course of time is entirely attributed to the change of the equilibrium vapour pressure (C_e).

The mass transfer coefficient which basically only depends on the nature of the surface, does not change significantly in the course of time.

The variation shown in this example in the mass transfer coefficient can be regarded as normal.

Table V for instance, gives total other values.

Example 3 : differentiation

In this case for k_{og} a value of about 5 x 10^{-4} m/s is found. Table V shows that it is not necessary to examine large sizes of particle board, but that samples of 0 by 15 cm usually are sufficiently representative.

Table V. Results of Six samples from the Same Board

sample	1	2	3	4	5	6
$\emptyset_g/A \times 10^3$ m/s			CH_2O	(ug/m^3)		
1.49	194	200	176	183	160	130
1.20	230	242	206	215	197	157
0.77	306	327	276	282	267	213
0.41	419	440	377	406	370	330
$0(=C_e)$	680	700	700	670	680	630
$k_{og}(10^4)$ m/s	5.6	5.7	5.2	5.0	4.1	3.5

Spread over the total width of the particleboard concerned, six samples have been examined. It is seen that the method allows differentiations over the surface. For instance, sample 6, which was taken from the edge, deviates from the other samples. Yet the difference is not so great that the particleboard as a whole would be misjudged.

Example 4 : ammonia treatment

The effect on particleboard of an ammonia treatment can also be shown using this testing method. In figure 8 again the ideal mixing model is applied. Notice that the line with the lowest emission is the one on the top. The reason is that the reciprocal values and not the steady state formaldehyde concentrations as such, are plotted. Here the slope is different as well.

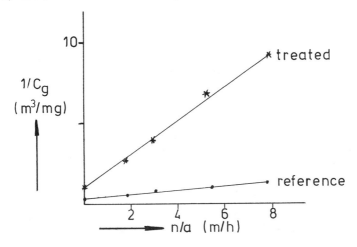

Figure 8. Treatment of particleboard with NH_3 gas.

This Figure and Table VI show that the difference concerns only
the equilibrium value. The effect of the treatment with ammonia is
that the equilibrium concentration is reduced drastically, while
as expected the mass transfer coefficient is not affected. So here
it is shown again that the slope as such does not have any
signification.

Table VI. Calculated Emission Parameters with NH_3 Treatment

		reference	NH_3
C_e	mg/m^3	8.1	0.87
$k_{og} \times 10^4$	m/s	3.0	3.0

Example 5 : other treatments

In Table VII some examples of treatments, also with an effect on
the mass transfer coefficient are shown. Four samples of the same
board are involved.

Table VII. Change of Formaldehyde Emission Parameters After
            Some Treatments.

		reference	24h/105°C	treated H_2O	soda
		1	2	3	4
C_e	mg/m^3	1.16	0.30	0.40	0.63
$k_{og} \times 10^4$	m/s	4.0	3.7	9.9	10.3
perforator	mg/100g	48	14	42	42
moisture	%	9.0	8.3	9.1	9.1

The first one was a reference sample.
The second one was dried at 105°C for a period of 24 hours.
The third one was "painted" with water in an amount of 135 g/m^2
and the last one was analogously treated with a diluted (20%) soda
solution.

After conditioning the moisture content of the boards was almost
the same as the original content, with exception of the dried
particleboard. The formaldehyde parameters of the treated samples
appeared to have changed very much.
    As a result of the treatment with water, the mass transfer
coefficient has increased in both cases.
    The equilibrium values of the treated samples had greatly
decreased as compared with the reference sample. The application
here of a soda solution had no favourable effect as compared with
the treatment with water only.

To illustrate that the relation with the perforator values (standard formaldehyde emission method, EN 120) is very poor, these values are given in the Table as well.

Example 6 : veneering

Figure 9 illustrates the effect of veneering on formaldehyde emission of particleboard. For the veneering the same type of resin was used as in the production of the particleboard. Pressing conditions are not comparable. Veneering has increased the equilibrium value a little, from 0.48 to 0.56 mg/m^3. The mass transfer coefficient however, decreased very much. The mass transfer resistance shows an increase from 2,400 sec/m to 11,000 sec/m. In the case at issue, the formaldehyde concentration, at a loading factor of 1 m^2/m^3 of the veneered particleboard, is below that of the bare particleboard, only at a ventilation rate in excess of 0.2 per hour.

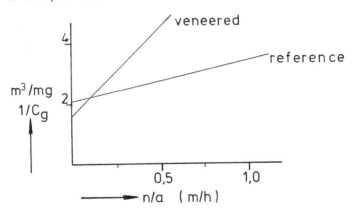

Figure 9. The effect of veneering.

BOARD COMBINATIONS

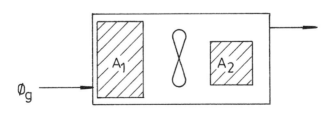

Figure 10. Board Combinations.

Sometimes several types of board are used in one environment.
(Figure 10) Assume that the environment is not ventilated and that
two types of particleboard are used. The equilibrium vapour
pressure of the two different boards generally are not the same.
As soon as the formaldehyde concentration in the air becomes
greater than the equilibrium concentration of one of the two
boards, this board will start to absorb formaldehyde instead of
emitting it. (For deduction of the mathematical equations, see
appendix 2.)

That this actually happens, can be demonstrated by placing
two different boards in a closed circuit with two burettes in
series, as shown in Figure 11.

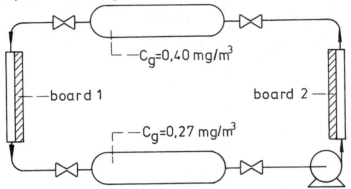

$-C_g = 0,40 \, mg/m^3$

$-$board 1   board 2$-$

$-C_g = 0,27 \, mg/m^3$

Figure 11. Two different board samples in a closed circuit.

After a few hours of circulating, different steady state
concentrations are in fact found in the two burettes. In other
words, one particleboard continually absorbs formaldehyde from the
other. In this case particleboard 1 absorbs formaldehyde from
particleboard 2. Table VIII shows the formaldehyde emission
parameters of the two boards. Especially the equilibrium values
are different, the mass transfer coefficients do not differ much.

Table VIII. FH Emission Parameters of the Boards of Figure 11.

	board 1	board 2
$k_{og} \times 10^4$ (m/s)	5.7	4.7
C_e (mg/m^3)	0.10	1.06
a (m^2/m^3)	0.5	0.5

The expected concentrations when both the particleboards are
placed in the same environment, are given in Table IX.

From the mass balance the concentration can be calculated as a function of the ventilation rate:

$$C_g = \frac{k_1.a_1.Ce_1 + k_2.a_2.Ce_2}{k_1.a_1 + k_2.a_2 + n}$$

in which a_1 = load factor (m^2/m^3) of particleboard 1
and a_2 = load factor (m^2/m^3) of particleboard 2   (see also
and   n   = ventilation rate (l/s)   appendix 2)

At ventilation rate zero, which means that there is no ventilation, there is not an equilibrium situation, but rather a stationary one.

Table IX.   Results of the Boards mentioned in Table VIII

vent. rate	separate + circ.		comb. + circ.	
	a_1=0.5	a_1=0	$a_1 + a_2$ = 0.5 + 0.5 = 1	
$n(h^{-1})$	a_2=0	a_2=0.5	board 1 + board 2	
			calc.	determ.
0	0.10	1.06	0.54	-
0.5	0.06	0.67	0.42	0.43
1.0	0.045	0.49	0.35	0.40
2.0	0.03	0.32	0.25	0.28
4.0	0.02	0.19	0.17	0.18

As can be seen from the results in Table IX, the overall formaldehyde concentration is not determined by the sum of the two concentrations, nor by the particleboard showing the highest release. For example at a ventilation rate of $0.5\ h^{-1}$ particleboard 2 with a loading factor of $0.5\ m^2/m^3$ gives a formaldehyde concentration of $0.67\ mg/m^3$. After addition of particleboard 1, with an extra loading factor of $0.5\ m^2/m^3$ (totally also $1\ m^2/m^3$), the calculated concentration based on the parameters, is $0.42\ mg/m^3$. That means lower as in the case of only board 2, even with halve the loading factor. The predictions on the basis of the theory agree with the values measured.

SUMMARY

The formaldehyde release of particleboard can, as far as the particleboard is concerned, be described by two characteristic parameters. The equation is:

$$1/C_g = 1/C_e + (1/k_{og}.C_e) \cdot n/a$$

In order to check whether the measuring system chosen, is suitable, the equilibrium value of the particleboard as such, is to be measured. (The intercept on the Y-axis has to be in accordance with the measurement of the C_e (equilibrium) value.)

In translating to formaldehyde concentrations in living
environments, an extra mass transfer resistance in the gas phase
must be taken into account. To give a quantitative impression, a
formulae is added here that could be used for practical purposes:

$$1/k_p \quad = \quad 1/k_{og} \qquad + \qquad 1/k_a$$

practice   board   air

approximately $1/k_a$ = 8600/3600 . a/n (s/m)

In principle this can only be done for an imaginary practical
living environment. In practice, many situations are more or less
approaching this imaginary situation.
    Provided that the characteristic parameters are known, the
formaldehyde concentration for combinations of boards, can be
calculated as well :

$$C_g = \frac{k_1.a_1.Ce_1 + k_2.a_2.Ce_2}{k_1.a_1 + k_2.a_2 + n}$$

Neither the simple sum of the concentrations nor the worst
particleboard is decisive.

Literature Cited

1. Hoetjer, J.J.; "Experiences with Measurements and
   Analytical Method for the Determination of the Formaldehyde
   Emission from Chipboard related to the Concentration in Living
   Environments"; Methanol Chemie Nederland vof, Delfzijl (1982)
2. Myers, G.E.; Nagaoka, M. Forest Prod.J. 1981, 31(7), 39-44.
3. Hoetjer, J.J.; Holz als Roh- und Werkstoff 1981, 39(9),
   391-393.

APPENDIX 1

Derivation of the mathematical equation for the formaldehyde
concentraction as a function of time in an enclosed space without
ventilation.

ASSUMPTIONS

By mixing the air, the formaldehyde concentration is homogeneous
with exception for a boundary layer, very close to the board.
    The amount of formaldehyde per unit of time emitted to the
air, is proportional to the installed surface (A) and the
concentration gradient $(C_e - C_g)$ with k_{og} (mass transfer
coefficient) as the proportional coefficient.

$\emptyset'_{FH} = k_{og} \cdot A \cdot (C_e - C_g) \quad (mg/s)$

C_g = (actual) formaldehyde concentration $\quad (mg/m^3)$
C_e = equilibrium concentration of the concerned board (mg/m^3)
A = surface $\quad (m^2)$
V = volume of the enclosed space $\quad (m^3)$
a = A/V specific area (loading factor) $\quad (m^{-1})$
t = time $\quad (sec.)$

The amount of formaldehyde ($\emptyset'_{FH} \cdot dt$) emitted to the air gives an increase of the formaldehyde concentration ($d\,C_g$) of the concerned volume of air (V).

$\emptyset'_{FH} \cdot dt = V \cdot d\,C_g; \qquad \emptyset'_{FH} = V \cdot \dfrac{d\,C_g}{dt}$

With the margin conditon at $t = 0; \quad C_g = 0$

The solution is: $\ln \dfrac{(C_e - C_g)}{C_e} = -k_{og} \cdot \dfrac{A}{V} \cdot t = -k_{og} \cdot a \cdot t$

APPENDIX 2

Deduction of the mathematical equation from the mass balance.
(ideal mixing model)

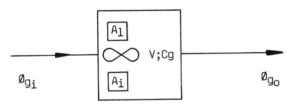

Emission of formaldehyde from the boards A_1, A_2 --- A_i per time is $\emptyset'_{FH_1}$, $\emptyset'_{FH_2}$, $\emptyset'_{FH_i}$ (see appendix 1).

$\emptyset'_{FH_1} = k_{og_1} \cdot A_1 \, (C_{e_1} - C_g)$

$\emptyset'_{FH_2} = k_{og_2} \cdot A_2 \, (C_{e2} - C_g)$

$\emptyset'_{FH_i} = k_{og_i} \cdot A_i \, (C_{ei} - C_g)$

NOTE: If $C_g > C_e$; $\emptyset'_{FH}$ is negatieve (the board is absorbing formaldehyde).

The incoming airflow is equal to the outgoing airflow (m^3/s). ($\emptyset_{g_i} = \emptyset_{g_o} = \emptyset_g$)

In a stationary situation the sum of the emitted amounts of formaldehyde is equal to the amount of formaldehyde, that is leaving the concerned room by ventilation with the air flow. This amount of formaldehyde is $\emptyset_g \cdot C_g$ (mg/s).

Equation: $\emptyset_g \cdot C_g = k_{og_1} \cdot A_1 (C_{e_1} - C_g) + k_{og_2} \cdot A_2 (C_{e_2} - C_g)$ --

$+ k_{og_i} \cdot A_i (C_{e_i} - C_g)$.

With $\emptyset_g/V = n$ (ventilation rate) (s^{-1})
and $A/V = a$ (specific area) (m^{-1}).

$$C_g = \frac{k_{og_1} \, a_1 \, C_{e_1} + k_{og_2} \, a_2 \, C_{e_2} \text{ --- } + k_{og_i} \, a_i \, C_{e_i}}{n + k_{og_1} \, a_1 + k_{og_2} \, a_2 \text{ --- } + k_{og_i} \, a_i}$$

In case of only one (separate) board the equation can be written as

$$1/C_g = 1/C_e + \frac{n}{k_{og} \cdot a \cdot C_e}$$

This gives a straight line if $1/C_g$ (m^3/mg) is plotted against n, $n/a = \emptyset_g/A$, $\emptyset_g$ or $n/a.C_e$.

To assure right application of the model the value of C_e (C_g at ventilation = 0) should be detected separately.

APPENDIX 3

Example of calculation with one kind of boardsurface.

ASSUMPTIONS

Specific area (loading grade) = 0,75 m^2/m^3 (E1 board).
C_e = 0,2 mg/m^3
k_{og} = 4 x 10^{-4} m/s

Expected formaldehyde concentration in a (climate) room with intensive cirulation at a rate of ventilation 1 h^{-1} = 1/3600 s^{-1}.

$1/C_g$ = 1/0,2 + 1/(3600 x 0,75 x 4 x 10^{-4} x 0,2) = 9,63 m^3/mg.

C_g = 104 ug.

Expected concentration at a rate of ventilation 0,5 h^{-1} = 0,5/3600 s^{-1}.

$1/C_g$ = 1/0,2 + 0,5/(3600 x 0,75 x 4 x 10^{-4} x 0,2) = 7,3 m^3/mg

C_g = 137 ug.

Estimation of the expected formaldehyde concentration under practical circumstances. At ventilation rate 1 h^{-1}.
Expected additional mass transfer resistance of the ambient air $(1/k_a)$

$1/k_a$ = a/n x 8600* = 0,75 x 8600 = 6450 sec/m.

Total mass transfer resistance $1/k_p$ = $1/k_{og}$ + $1/k_a$ = 8950 s/m.
Expected concentration:

$1/C_g$ = 1/0,2 + 8950/(3600 x 0,75 x 0,2) = 21,6 m^3/mg.

C_g = 0,046 mg/m^3.

At ventilation 0,5 h^{-1}.

Expected additional resistance : 0,75/0,5 x 8600 = 12900 s/m.

Total resistance : 12900 + 2500 = 15400 (s/m).

$1/C_g$ = 1/0,2 + (15400 x 0,5)/(3600 x 0,75 x 0,2) = 19,3 m^3/mg.

C_g = 52 ug/m^3.

* = arbitrarily value

APPENDIX 4

Example of calculation with two kinds of boardsurfaces.

ASSUMPTIONS

Specific area a_1 (f.i. coated panels) 0,75 m^2/m^3.
Specific area a_2 (f.i. uncoated sides total open to the air) 0,03 m^2/m^3.

a_1 = 0,75 m^2/m^3 k_{og_1} = 5 x 10^{-6} m/s C_{e_1} = 0,2 mg/m^3
a_2 = 0,03 m^2/m^3 k_{og_2} = 5 x 10^{-4} m/s C_{e_2} = 1,5 mg/m^3

Expected formaldehyde concentration in an intensive circulated room at a ventilation rate 1 h^{-1} = 1/3600 s^{-1}.

$$C_g = \frac{5 \times 10^{-6} \times 0,2 \times 0,75 + 5 \times 10^{-4} + 0,03 \times 1,5}{5 \times 10^{-6} \times 0,75 + 5 \times 10^{-4} \times 0,03 + 1/3600} = 0,078 \text{ mg/m3}$$

C_g = 78 ug.

At a ventilation rate of 0,5 h^{-1} C_g = 147 ug/m^3.

Estimation of the expected concentration under practical
conditions, at a ventilation rate 1 h^{-1}.
Calculated total mass transfer resistance $(1/k_{p_1})$ of installed
a_1 m^2/m^3.

$1/k_{a_1}$ = 0,75 x 8600 = 6450 s/m

$1/k_{og_2}$ = 1/5 x 10^{-6} = 2 x 10^5

$1/k_{p_1}$ = $1/k_{a_1}$ + $1/k_{og_1}$ = 206450 k_{p_1} = 4,8 x 10^{-6} m/s

Calculated total mass transfer resistance $(1/k_{p_2})$ of installed a_2
m^2/m^3.

$1/k_{a_2}$ = 0,03 x 8600 = 258 s/m

$1/k_{og_2}$ = 2000

$1/k_{p_2}$ = 2258 k_{p_2} = 4,4 x 10^{-4} m/s

Calculated C_g = 70 ug/m^3.

At a ventilation rate 0,5 $^{-1}$ calculated total resistance of a_1:

$1/k_{p_1}$ = $1/k_{og_1}$ + $1/k_{a_1}$ = 1/5 x 10^{-6} + (0,75/0,5) x 8600 =

212900 s/m

k_{p_1} = 4,7 x 10^{-6} m/s

Calculated total resistance of a_2:

$1/k_{p_2}$ = $1/k_{og_2}$ + $1/k_{a_2}$ = 1/5 x 10^{-4} + (0,03/0,5) x 8600 =

2516 s/m.

k_{p_2} = 4,0 x 10^{-4} m/s

$$C_g = \frac{4,7 \times 10^{-6} \times 0,2 \times 0,75 + 4,0 \times 10^{-4} \times 0,03 \times 1,5}{4,7 \times 10^{-6} \times 0,75 + 4,0 \times 10^{-4} \times 0,03 + 0,5/3600} = 121 \text{ ug/m3}$$

RECEIVED January 14, 1986

Measurements of Formaldehyde Release
from Building Materials in a Ventilated Test Chamber

Hans N. O. Gustafsson

National Testing Institute, Box 857, S-501 15, Boras, Sweden

Formaldehyde as a pollutant in the indoor air is
usually connected with the use of formaldehyde based
resins in e.g. building materials and in furniture.
This article presents measurements of the formaldehyde emission from various products containing urea-
formaldehyde (UF) or phenol-formaldehyde (PF) resins.
The emission from all test objects have been measured
in a ventilated test chamber at the standardized
testing atmosphere 23 °C, 50 % RH according to the
International Organization for Standardization (ISO).
The emission from woodbased panels and other
materials have been measured at a loading factor of
1.0 m^2/m^3 and at an air change rate of 1.0 h^{-1}.
The values of the test variables are in agreement
within the work of the European Organization for
Standardization (CEN).
  Woodbased panels have also been tested with
the perforator method. This method is European Norm
according to CEN and gives an estimate of the
extractable content of formaldehyde for especially
particle boards. Formaldehyde release has also been
investigated for different kind of pieces of furniture exposed in area to volume proportions in which
they can be found in a small room.

In Denmark, Finland, Norway, Sweden and in West Germany the content
of formaldehyde in woodbased panels are regulated by perforator
values. In Denmark and West Germany these rules furthermore are
based upon requirements of the formaldehyde emission to the air in
ventilated test chambers. The regulations in Sweden include at the
moment only UF-bonded particle boards. The boards should not exceed
a perforatorvalue of 40 mg "free formaldehyde" per 100 gram dry
board.

0097–6156/86/0316–0145$06.00/0
© 1986 American Chemical Society

In the future the Swedish formaldehyde rules may include
other UF-bonded products as MDF-boards and the requirements also may
be formulated as emission rates. On the behalf of the National Board
of Physical Planning and Building, the Swedish National Testing
Institute has performed a study on the emission from products bonded
with formaldehyde based resin. The measurements have been performed
in a ventilated test chamber at standardized climate in agreement
within the work of the European Organization for Standardization,
CEN. 16 West European countries are represented in CEN.

The aim of this study was to compare

- the emission rate from different woodbased panels and other
  materials
- the emission rate with the perforator value
- the contribution of formaldehyde from different pieces of
  furniture to the total level of formaldehyde in a small room.

The study does not include comparison of different types of
diffussion barriers.

Complete results with a closer description of the test
objects and a review of official testing methods in the Nordic
countries and West Germany are presented in a technical report from
the Swedish National Testing Institute (1)

Materials

The tested products were bonded with formaldehyde containing resin
and used indoors. Most of the woodbased panels and other materials
were manufactured during 1984. The panels were not coated. If
nothing else is stated the test objects were manufactured with
UF-resin and of Swedish origin. As the most common UF-bonded
material the particleboards dominated the investigation. The selected boards included both ordinary UF-bonded (V-20) and moisture
resistant boards, (V-313) and were received from eleven factories.

The nine MDF-boards that were tested were from five different manufacturers in Europe and have been commercially available in
Sweden. One of the boards was moisture resistant and another flame
resistant. Two of the boards were treated with formaldehyde reducing
agents.

Other woodbased panels as UF-bonded plywood, blockboard,
PF-bonded plywood and hardboard/fibre building board have also been
tested. Emission tests have even been performed for materials such
as UF-foam (UFFI), mineral wool, plasterboard and furniture foil.
The UF-foam was manufactured by a licenced contractor in 1979 and
had never been installed in a building.

Test furniture, decoration panels and laminated parquet
flooring were purchased during 1984.

Methods

Emission. As the emission varies considerably with temperature and
relative humidity of the air (2) it is necessary that the test is
performed at constant climate. Our test conditions were in agreement
with the tentative mehod of CEN (3).

The CEN method is based upon the assumption that the size and shape of the testing chamber does not influence the emission. During the testing the formaldehyde concentration in the chamber will rise and stabilize at a steady state concentration. At constant climate the steady-state concentration or emission rate from the test object depends on the relation between the loading factor and the air change rate. Good air circulation in the chamber is also essential (4).

Formaldehyde emission was measured at 23°C and 50 % RH in a ventilated test chamber of 1.0 m^3, the testing climate recommended by the International Organization of Standardization (ISO) (5). The exposed area of woodbased panels and other materials were 1.0 m^2. Thus the loading factor in the chamber is 1.0 m^2/m^3. The air change rate was 1.0 airchange per hour. Since the emission from the edges of the board often are higher, the edges are consequently sealed with self-adhesive aluminium tape. For precondi- tioned test pieces the steady state concentration will be reached within a week.

Pieces of furniture have been tested in proportions in which they may be found e.g. in a small living room. The room is assumed to have a floor area of 7 m^2 with a height of 2.4 m. The air volume in this room is 17 m^3. The loading factor for the tes- ted floors is thus 7/17 = 0.4 m^2/m^3. Decorative panels have been tested at a loading factor of 1.0 m^2/m^3.

Each of the test chambers has an internal volume of 1.0 m^3 and consists of stainless steel, with the dimension 1 000 x 1 500 x 667 mm. The chambers are supplied with air of constant tem- perature (23 ± 0.5°C) and constant relative humidity (50 ± 3 % RH) from a conditioning plant. The background concentration of formaldehyde in the supplied air is regularly measured and is less than 0.02 ppm. The air exchange rate from the chambers is performed within ± 3 % by exhaust pumps. The extract air from the chambers is not recirculated. The leakage of air into the chambers have been measured to be less than 1 % at an air change rate of 5 per hour.

The concentration of formaldehyde in the chamber air is determined spectrophotometrically after sampling in bottles. Chromotropic acid (6) or acetyl-acetone (7) were used as reagents. Acetylacetone reacts more specific with formaldehyde but the reac- tion requires a higher temperature to be quantitive.

Extraction with toluene.   The extractable content of free formaldehyde in woodbased panels have been estimated with the perforator method. This method is an European Norm (8). With this method the test pieces (25 x 25 mm) are boiled in toulene in 2 h. The toulene is condensed continously and brought in contact with water, which is titrated iodometrically.The perforator apparatus is made up of several different glass parts.

Results

All data presented in this paper constitutes steady-state values that are the average of at least 3 measurements. The relative stan- dard deviation of the presented steady-state values is about 5 %.

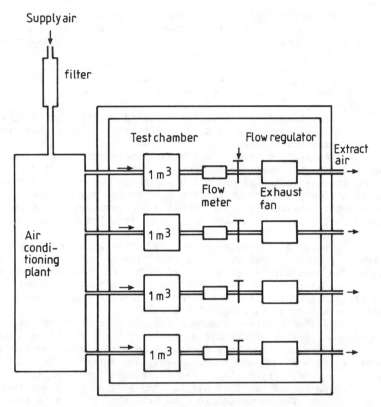

Figure 1. The air conditioning plant supplies each
chamber with air of constant temperature and constant
relative humidity.

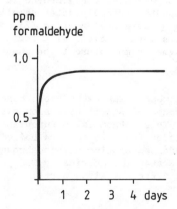

Figure 2. During the testing the concentration of
formaldehyde in the chamber will rise and stabilize
at a steady state level.

Woodbased panels and other materials. At an air change rate of 1,0 per hour the given steady-state concentration corresponds to an emission rate with an equal number of ppm formaldehyde/m^2 x h. The emission from the tested particle boards and MDF-boards are given in figure 3 and figure 4 respectively. The emission from boards except particle boards are presented in table I.

Table I	Steady-state concentration ppm formaldehyde 23°C, 50 % RH air change rate 1,0 h^{-1} 1,0 m^2/m^3	Perforator value mg formaldehyde/ 100 g dry board	Treatment/ Origin
Blockboard	0,08	13	
MDF-boards	1,8	63	
	0,9	50	
	2,0	71	
	1,7	70	moisture resistant
	3,2	125	flame resistant
	1,7	86	
	0,20	27	posttreated with NH_3
	0,31	23	
	0,13	10	posttreated with $(NH_4)_2$ CO_3
Plywood	0,02	4	PF-bonded
	0,22	34	A-70
	0,67	27	Far East
Hardboard/ Fiber building board	0,02	3	PF-bonded
Pure wood	<0,02	–	
Plasterboard	<0,02	–	
UF-foam	0,23	–	
Mineral wool	0,02	–	
Furniture foil	0,28	–	

Pieces of furniture. During the testing the air change rate has been 0,5 h^{-1}, which is close to practice in the Nordic countries. The various types of pieces of furniture has been tested at different area to volume proportions as in actual conditions. The emission from pieces of furniture are presented in table II.

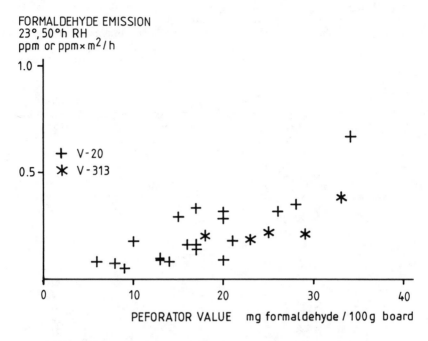

Figure 3. Formaldehyde emission versus perforatorvalue for particle boards. The emission could be
expressed either as a steady state concentration (ppm)
or an emission rate (ppm x m²/h).

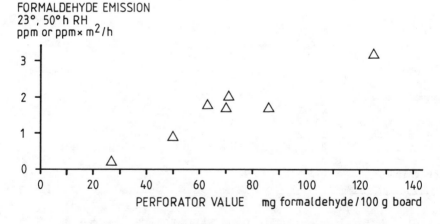

Figure 4. Formaldehyde emission versus perforatorvalue for MDF-boards. Observe the different scale on
the Y-axis compared with figure 3.

Table II

Test object to be placed in a room (7 m²) with an air change rate of 0,5 h⁻¹	Steady state concentration ppm	Remarks
Parquet floor I	0,11 0,06	re-teasted after 4 months
Parquet floor II	0,02	
Decorative paneling	1,0	Lauan-type
Decorative paneling	0,7	coated with monstered paper
Front of cupboard	0,02	veneered particleboard
Front of cupboard	0,06	painted MDF-board

Discussion

Woodbased panels and other materials. The UF-bonded mineral wool insulation releases only barely measureable quantities of formalde-hyde. This stems probably on the frequent addition of urea in the manufacturing process.

PF-bonded materials as plywood and hardboard/fibre building board also release only very small quantities of formaldehyde. Low release from PF-bonded plywood have also been shown with another method (9).

All woodbased panels were also investigated with the per-forator method. Even though, strictly, this European Norm is applicable only for particle boards, the method is used, in praxis, even for other non-coated UF-bonded boards. There is no linear rela-tion between the emission and the perforator value for e.g. particle boards, as can be seen in Figure 3.

Particle boards produced at the same factory however normally have a good correlation between emission and perforator value. The official Danish and West German requirements are based on this fact.

Moisture resistant boards are manufactured of UF with some melamine added (MUF-bonded). If these boards are excluded from the calculations the correlation factor between the perforator values and emission values increases from r=0,76 to r=0,82.

Most of the tested MDF-boards release large quantities of formaldehyde. Boards with lowest emission have been posttreated with formaldehyde-reducing agents as gaseous NH_3 or $(NH_3)_2 CO_3$, which react with the formaldehyde.

As the weight content of UF-resin are both about 10% (counted as dry weight of the resin per dry wood) in MDF-boards and particle boards, it is not possible to explain the different emission rates.

While the perforator method also can be used for production control of MDF-boards it is questionable weather the method is feasable for plywood and other laminated wood panels. The two tested UF-bonded plywood boards e.g. although equal perforator values shows large difference in emission.

Furnishing. The formaldehyde level in a room at actual conditions depends on several factors, and is not an arithmetical sum of various sources (10), (11). In order to estimate the contribution of formaldehyde emission from single pieces of furniture the test objects have been exposed in area to air volume proportions to which they can be found in a small room or a kitchen. The assumption that the formaldehyde level in the chamber and in the actual room is the same, is based on a theoretical model originally developed for particle boards (4). At constant climate the emission from a test object is determined of the relation between the loading factor and the air change rate.

The results show that the emission from UF-bonded decorative paneling could rise to high levels in a room.

The rapid decrease of emission in one of the floorings seems to be due to the hardening of the acid curing laquer layer and not to the ageing of the UF-resin in the laminated construction.

Acknowledgment

The author wish to thank Mr Birger Sundin, AB Casco, for valuable discussions during the project.

This research was sponsored by the Swedish Council for Building Research.

Literature Cited

1. Gustafsson, H.N.O., Isaksson, I, Muameleci, E:
"FORMALDEHYD TILL INOMHUSLUFT - Mätningar i ventilerad kammare av byggmaterial och inredningar vid standardklimat": Statens Provnings-anstalt: Borås, Sweden, Teknisk rapport 1985:29 (In Swedish, 85 p)

2. Andersen, I, Lundquist, G.R., Mölhave, L: "Indoor air pollution due to chipboard used as a construction material. Atmos Envir 1975,9,1112-1127.

3. "Particle boards - Determination of formaldehyde emission under specified conditions" Method called: Formaldehyde Emission Method; European Committee for Standardization, CEN: 1984; CEN report, CR 213

4. Hoetjer, J.J.: "Experiences with measurements and analytic method for the determination of the formaldehyde emission from chipboard related to the concentration in living environments". Methanol Chemie Nederland v o f: Delfzijl 1982

5. Standard atmospheres for conditioning and/or testing specifications; International Organization for Standardization, 1 ed. 1976; ISO 554

6. "Formaldehyde in Air, Method No: P&CAM 125". Manual of Analythical Methods; 2 ed. NIOSH: Cincinnati: 1977.

7. Menzel, W. Marutzky, R. und Mehlhorn, L. "Formaldehyd – Messmetho- den" WKI-Bericht Nr. 13 Frauenhofer-Institut für Holzforschung, West Germany: 1981

8. "Particle boards – Determination of formaldehyde content extraction Method called: Perforator Method; European Committée for Standardization, CEN;EN 120: 1984

9. Meyer, C. B. "Formaldehyde Release From Phenolic Bonded Wood Panels". American Plywood Association: 1981

10. Myers, G.E. Formaldehyde dynamic air contamination by hardwood plywood: effects of several variables and board treatments. For. Prod. J. 1982, 32,4,20-25

11. Godish, T, Kanyer, B. Formaldehyde source interaction studies. For. Prod. J. 1985,35,4,13-17

RECEIVED January 14, 1986

13

Large-Scale Test Chamber Methodology for Urea–Formaldehyde Bonded Wood Products

L. R. Newton, W. H. Anderson, H. S. Lagroon, and K. A. Stephens

Georgia-Pacific Corporation, L. F. Bornstein Research Laboratory, Thermoset Resin Division, 2883 Miller Road, Decatur, GA 30035

The U.S.Department for Housing and Urban Development's rule 3280.308 established formaldehyde emission standards for particleboard and hardwood plywood paneling used in mobile homes. These standards took effect February 11, 1985. The certification program under this rule requires each manufacturer to develop a quality control in-plant testing program that relates to tests conducted in a large scale environmental chamber.

This paper presents Georgia-Pacific's and other investigators' experience with various aspects of large environmental chamber design and operation. Experimental data and observations are presented in such topics as: 1.) Common formaldehyde air test methods; 2.) Formaldehdye generation and recovery studies; 3.) Air exchange measurement techniques; 4.) Preconditioning of test boards; 5.) Temperature effect on chamber formaldehyde concentrations; 6.) Relationship of popular quality control test methods to the large chamber; 7.) Loading, air exchange rate, and wood product combination effects on chamber formaldehyde concentrations; 8.) Chamber Round Robin studies between Georgia-Pacific's chamber and other outside lab chambers; 9.) Chamber concentrations and its relationship to actual field measurements.

The recent implementation of the U.S. Department of Housing and Urban Development (H.U.D.) formaldehyde emission standards for particleboard and hardwood plywood paneling used in mobile homes is the first enforced government formaldehyde standard in the world (1). In Europe, there are voluntary formaldehyde product standards in several countries(2); however, there is no official government enforcement or verification program in place that those standards are being met(3). The HUD rule, which went into effect February 11, 1985, requires mobile home manufacturers to use particleboard and hardwood plywood paneling that do not exceed formaldehyde chamber concentrations of 0.3 ppm and 0.2 ppm, respectively, at specific product loadings.

0097-6156/86/0316-0154$09.50/0
© 1986 American Chemical Society

The certification program outlined by HUD requires that each
wood product manufacturer develop a quality control plan that will
provide a basis of relating its "in-plant" testing program to
quarterly tests conducted in large scale test chambers (24.1 cubic
meter minimum volume)(4). The "in-plant" testing program requires
the use of a quality control test (i.e. 2 Hour Desiccator(5),
Equilibrium Jar(6), 24 Hour Desiccator(7), etc.) that relates to the
large scale test chamber. The quality control test must be
sensitive and reliable enough to monitor day-to-day formaldehyde
emission variations in the wood product. Nationally recognized
testing laboratories provide the mechanism to certify wood products
to meet the HUD rule by approving written quality control plans,
perform routine in-plant inspections, conduct large chamber testing
on a quarterly basis, and spot check quality control testing. Thus,
the emphasis on large scale chamber tests and quality control
testing is the heart of the rule that assures formaldehyde emissions
from U-F resin bonded wood products are acceptable for use in mobile
homes. To date, the HUD Rule has had a tremendous impact on all
wood products bonded with U-F resins. Major distributors, home
manufacturers, and contractors are requesting HUD certified wood
products for use in conventional house and office construction.
Many wood manufacturers have responded by producing, advertising,
and certifying that their U-F bonded wood products meet HUD
formaldehyde standards.
    In this paper, we will present experimental formaldehyde
emission data obtained on a variety of U-F bonded wood products.
This data was gathered over a three to four year period from chamber
and various quality control tests conducted at the G.P.
L.F.Bornstein Research and Development Laboratory, Decatur, Georgia;
G.P. Laboratory, Sacramento, California; Georgia-Tech Research
Institute, Atlanta, Georgia; Hardwood Plywood Manufacturers
Association, Reston, Virginia; National Particleboard Association,
Gaithersburg, Maryland; and several major wood manufacturers.

Environmental Test Chamber

The first environmental test chamber located in the L.F. Bornstein
Research and Development Laboratory was constructed in January,
1981(8). A second chamber was installed beside the earlier chamber
in April, 1983. The chambers were constructed to simulate the
ambient indoor environmental conditions found in a mobile home.
Both chambers are 28.4 cubic meters in volume. This is about one-
fifth of the volume of a single-wide mobile home. Wood products are
loaded into the chamber at a given surface-to-chamber volume ratio
(m2/m3) based on product type. The temperature, relative humidity,
and air exchange rate are maintained at 25+0.5°C, 50+4% relative
humidity, and 0.50+.05 air changes per hour respectively. The wood
product remains in the chamber until a steady formaldehyde
concentration is obtained. To lessen the time the wood product is
in the chamber, the Hardwood Plywood Manufacturers Association and
the National Particleboard Association provide in their "Large Scale
Test Chamber Test Method" FTM-2 a seven day "preconditioning" period
outside the large test chamber at similar large chamber
environmental parameters. This "preconditioning" procedure occurs
prior to the wood product insertion into the large test chamber.

Chamber Design

Figure 1 is a top view sketch of our environmental test chamber.
The chamber is constructed of mill finished stucco embossed aluminum
with a 20 gauge heavy duty galvanized steel floor. The choice of an
all aluminum structure was based on the lack of reactivity of
formaldehyde with aluminum (9), observed adsorption of formaldehyde
on steel sheet metal(10), and the availability of a prefabricated
walk-in cooler.
     The internal dimensions of the chamber are 2.23 meters wide,
5.29 meters long, and 2.41 meters high. Allowing for internal
equipment volume of 0.06 cubic meters, the effective volume is a
little less than 28.4 cubic meters.
     As indicated in Figure 1, the chamber is equipped with epoxy
coated steel angle iron rack (Item 5) to support the wood samples
(1.2 meter X 1.2 meter) in a vertical position. Based on
cooperative tests conducted by Georgia-Tech and Georgia-Pacific,
orientation of the board samples in a horizontal or vertical
position does not seem to affect chamber concentration provided
there is sufficent distance between boards to allow reasonable air
flow across the board surface. We recommend a minimum of 20
centimeters between paneling surfaces and 30 centimeters between
particleboard board surfaces for chambers designed like ours. In
our chamber tests, the back and front of particleboard and paneling
are exposed to the chamber's interior. The air flow observed across
boards in our chamber range between 6 to 15 meters per minute.
Based on our and other researchers' experience, the minimum air
flow across the board should be between 1.5 to 6 meters per
minute(11). Current chamber research at the National Bureau of
Standards (N.B.S) sponsored by the Consumer Products Safety
Commission on formaldehdye emissitivity from pressed wood products
is being conducted at 1.5 meters per minute face velocity(12).
N.B.S. researchers believe the 1.5 meters per minute face velocity
is realistic of actual air flow in a dwelling. Our chamber studies
indicate that face velocities become an important factor in
determining final chamber formaldehyde concentration whenever the
board is classed as a high emitter. High face velocities for high
emitters appear to promote higher chamber concentrations. However,
high face velocities across low emission boards do not appear to
appreciably affect chamber concentrations.
     The G.P. chamber is equipped with an air cooler of about 5500
BTU size (Item 3) is located about 1.8 meters off the floor. An
evaporator control valve (Item 2) on the refrigerate line allows
temperature control of the air cooler condenser coils. Temperature
of the coils is maintained just above the dew point for 50% relative
humidity. An electric baseboard heater (Item 6) is equipped with a
hydrastatic thermostat control (Item 11). The base board heater is
placed near the floor and opposite of the chamber door. A
humidifier (Item 4) is located approximately 1.8 meters above the
floor and just to the right of the air cooler. The nozzle of the
humidifier is pointed slightly toward the back of the chamber. A
humidistat (Item 10) is centered between the floor and end walls on
the same side as the humidifier. A strategically located recording
hygrothermograph is used to monitor both temperature and humidity.
We have found it is best to back up the hygrothermograph with a dial
hygrometer and thermometers.

To stabilize temperature, relative humidity, and formaldehyde concentrations within the chamber, we have found it necessary to have an air deflector (Item 13) placed between the back wall and a floor fan (Item 14) in such a way that the air flow from the floor fan is directed counter-current to the air flow movement from the air cooler's blower. Formaldehyde recovery studies, smoke stick evaluations, and formaldehyde determinations performed in several locations within the chamber have substantiated the efficiency of this mixing technique.

Two remote sampling probes of 0.635 centimeter I.D. TEFLON are located equal distance from each end wall and from each other. The probe inlet is located approximately 1.4 meters off the floor of the chamber. A third sample probe located in the 3.8 centimeter exhaust hole (Item 12) provides an occasional verification of mixing consistency within the chamber. Formaldehye measurements at all sample locations have always checked within the experimental precision of the analytical method (approximately 4% for a 60 liter air sample). Obviously, these probes provide a convenient way of sampling the air within the chamber without distrubing the established chamber equilibrium.

The fresh air make-up for the chamber is provided by a variable speed Roots blower with an automobile air filter placed ahead of the blower intake. (A number of laboratories have had success in utilizing less expensive cage blowers with valves placed in line to control air flow.) The air from the blower is passed through an air dryer to reduce moisture content to about 20 to 30% relative humidity. From here, the air is then passed through a bed of PURAFILL II Chemisorbant to reduce formaldehyde in the incoming air to levels less than 0.02 ppm(vol./vol.). A 1.27 centimeter critical orifice and ball valve (Item 9) are located just ahead of a Singer Diaphram Gas Meter, Model No. AL-800 (Item 8). The air exits the gas meter and enters the chamber through a 3.8 centimeter I.D. by 122 centimeter long PVC diffuser tube (Item 7).

The amount of air passing into the chamber is totalized by the diaphragm dry gas meter (Item 8). The air change per hour is computed by taking the difference of two gas meter readings and dividing by the chamber volume and time interval for the meter readings.

$$\text{ACPH} = \frac{\text{V2} - \text{V1}}{28.4 \text{ CU.METERS X } \Delta t} \tag{1}$$

Where: ACPH is air changes per hour
V2(m3) is the ending meter reading at time T_f
V1(m3) is the beginning meter reading at T_o
28.4 cu. meters is our chamber volume
Δt is the time interval in hours between
meter readings (T_f-T_o)

Accuracy of the diaphragm gas meter is verified against either a wet test meter or a Sierra Instruments 616 E-36 hot wire aneomemeter. On a yearly basis, a third party laboratory verifies chamber operation and air exchange rate measurements. The carbon monoxide decay is the method used to verify air exchange rate measurements(13). However, other researchers have reported using formaldehyde, propane, sulfur hexafluoride, and carbon dioxide as tracer gases to verify air exchange rates; based on their findings, it appears it may take some time for gas decay to stabilize in the chamber before consistent air change rates are observed (14). In a recent meeting of chamber operators, the general consensus was that in line totalizing gas meters were far more accurate than gas decay techniques. This consensus was based on the consistancy observed on a day-to-day basis with gas meters(15).

Chamber Protocol For Testing Wood Products

The H.U.D. rule refers to the " Large Scale Test Method for Determining Formaldehyde Emission from Wood Products" FTM-2 - 1983 (16). In this method, particleboard and hardwood plywood paneling are tested under the following conditions:

Table I. H.U.D. Chamber Test Conditions For U/F Bonded Wood Products

	Particleboard	Paneling
Loading (M2/M3)	0.43	0.95
Temperature (deg.C)	25+1	25+1
Relative Humidity	50+4%	50+4%
Air changes per hour	0.50+0.05	0.50+0.05

Formaldehyde Measurement Methods For Chamber & Field Concentrations

Ambient formaldehyde determinations taken during large scale test chamber studies and field investigations are based on two colorimetric analyses. The two methods are: a modification of NIOSH P&CAM 125 and the CEA 555 continuous formaldehyde monitor.

The modified NIOSH P&CAM 125 method utilizes two 30 mL midget impingers each containing 20 mL of 1% sodium bisulfite (NaHSO3) collection medium. The amount of collection medium is weighed into each impinger. With the impingers connected in series to a M.S.A. Fix-Flo personnel pump, air is bubbled through the impingers at a rate of 1 liter per minute for one hour. Pre- and post-calibration of the personnel pump is performed for each sample collection. The impingers are reweighed and adjusted to the original weight with fresh 1% sodium bisulfite collection solution. The total amount of solution required for this adjustment seldom exceeds 0.5 gram for both impingers. The scrubbing efficiency of the first impinger is 95.9% with relative standard deviation of 3.5%. Formaldehyde collected in 1% sodium bisulfite may be stored at room temperature with little or no loss in concentration for up to 1 month. Refrigerated samples can be held almost indefinitely. However, it is our practice to analyze all collected air samples within 24 hours after collection. Results are expressed in ppm (vol./vol.) formaldehyde.

A good substitute absorbing solution for the 1% sodium bisulfite solution is 0.1 N sodium hydroxide. Based on several years of testing, we have found the 0.1 N sodium hydroxide has the same scrubbing efficiency and analytical quality as the 1% sodium bisulfite absorbing medium. The Cannizzaro reaction has not been a factor in reducing the amount of formaldehyde in collected air samples held at room temperature for 2-3 days.

The CEA 555 continuous monitor(17) is used for real-time
monitoring of chamber and actual field survey formaldehyde
determinations. The monitor is a useful instrument in field surveys
because it is one method that provides real time formaldehyde
measurements which is useful in tracing and identifying usually high
formaldehyde sources. The monitor's analytical method is based on
the modified Schiff procedure developed by Lyles, Dowling and
Blanchard (18). Formaldehyde is absorbed in a sodium
tetrachloromercurate solution that contains a fixed quantity of
sodium sulfite. Acid bleached pararosaniline is added, and the
intensity of the resultant dye is measured at 500 nm. Both
formaldehyde in air and liquid standards can be analyzed.

We have conducted side by side tests using the CEA 555 Air
Monitor and the Modified NIOSH P&CAM 125 method in 19 actual field
surveys of conventional homes, mobile homes, and offices over a one
year period. The nineteen data points are graphically depicted in
Figure 2. As can be seen, there is an excellent correlation of the
Modified NIOSH P&CAM 125 to the CEA 555 Air Monitor method.

Chamber Formaldehyde Recovery Studies

Georgia Institute of Technology Studies. Georgia Institute of
Technology performed formaldehyde recovery studies in the Georgia-
Pacific Environmental Chamber while doing a research project (18).
Known concentrations of formaldehyde were achieved with a
formaldehyde generator designed by Dr. Jean Balmat, formerly of
DuPont. At this time, design information of this generator cannot be
released due to pending publication by Dr. Balmat and DuPont
personnel(19).

The GIT formaldehyde recovery studies in the chamber were
performed at three separate concentrations, 0.1 ppm, 0.4 ppm, and 2.5
ppm. Chamber operating conditions of 24 degrees Centigrade, 50% RH,
and 0.5 ACPH were maintained for each of these evaluations. A known
concentration of formaldehyde was introduced into the chamber using
the DuPont formaldehyde generator. Formaldehyde concentration in the
chamber was continuously monitored using the CEA 555 continuous air
monitor instrument. When the steady state level of formaldehyde was
reached, the chamber formaldehyde concentration was determined using
the modified P&CAM 125 method. Two measurements were made for each
concentration. Recoveries were considered excellent (> 92% for each
of the three concentrations). "Considering the experimental error of
the technique (estimated at 8%), the HCHO loss within this specific
large scale environmental chamber under the described conditions was
minimal" (20). Table II is a summary of the Georgia Tech first
recovery study:

Table II. Georgia Tech Chamber Formaldehyde Recovery Study #1
@ 25°C, 50% RH

Target HCHO Concentration (ppm HCHO)	Percent Recovery (%)	Air Change Rate (No./Hr.)
2.5	95.7	0.50
0.4	95.5	0.50
0.1	92.8	0.50

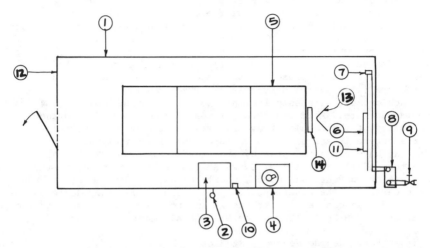

Figure 1. Top view sketch of test chamber.

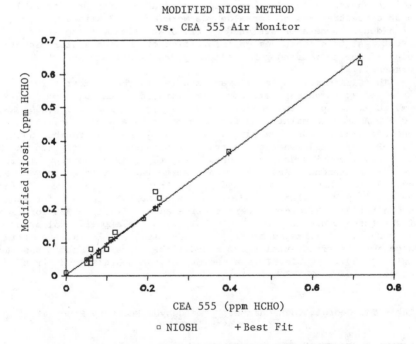

Figure 2. Modified NIOSH P&CAM125 vs CEA 555 air monitor.

A second formaldehyde recovery study by Georgia Tech in their
large scale test chamber agrees very well to study #1. The Georgia
Tech test chamber is modeled after the Georgia-Pacific chamber used
in study 1. In a to be released report (21), a summary of their
second study is as follows:

Table III. Georgia Tech Chamber Formaldehyde Recovery Study #2
@ 25°C, 50% RH

Target HCHO Concentration (ppm HCHO)	Percent Recovery (%)	Air Change Rate (No./Hr)
0.10	92.5	0.53
0.40	92.0	0.53
0.10	93.2	1.10
0.40	90.6	1.10

Georgia-Pacific Recovery Studies. For us to perform our own recovery
study, we refined and developed a syringe pump method for generating
formaldehyde concentrations within our large scale test chamber. This
method was originally created by Mr. Bill Lehnman of Weyerhaeuser,
Tacoma, Washington (22).

This simple approach involves the introduction of formaldehyde
into the test chamber at a known concentration based on theoretical
calculations involving chamber volume, air change rate, and syringe
pump delivery rates. Figures 3 & 4 are drawings of the syringe
pump assembly and evaporator oven, respectively.

The syringe is mounted in the syringe pump apparatus which is
positioned in the test chamber in a central location. Prior to
testing, the light bulb which heats the evaporator oven is turned on
approximately 12 hours before formaldehyde is generated. Heat from
the light bulb increases the chamber temperature by about 1°C. The
generator is placed in the chamber so there is adequate dispersion of
the generated formaldehyde gas.

The oven is constructed out of aluminum foil as shown in Figure
4. The syringe needle is inserted into the oven approximately 7.62
centimeters above the 100 watt light bulb. Once the light bulb oven
is up to temperature the syringe pump is activated causing drops to
fall from the syringe needle. These drops must not be allowed to
fall on the heat source until the 0.50 mL stock solution "SPIKE" is
injected onto the heat source and vaporized. Once the "SPIKE" is
vaporized the syringe drops can then be allowed to fall on the heat
source. The "SPIKE" will push the formaldehyde concentration to the
predetermined target concentration (for example 0.40ppm). The
syringe pump will maintain the target concentration at the given air
change rate until the syringe is empty, approximately 8 hours.

The following Table IV gives recovery efficiencies we observed
using our generator system for target concentrations of 0.30ppm and
0.40ppm. Eight determinations were made per target concentration at
0.50 air change rate, 25+1°C, and 50% relative humidity.

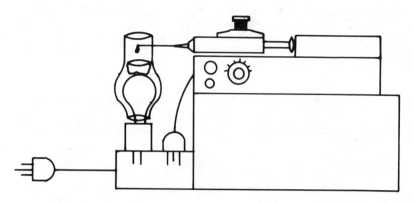

Figure 3. Syringe pump formaldehyde generator.

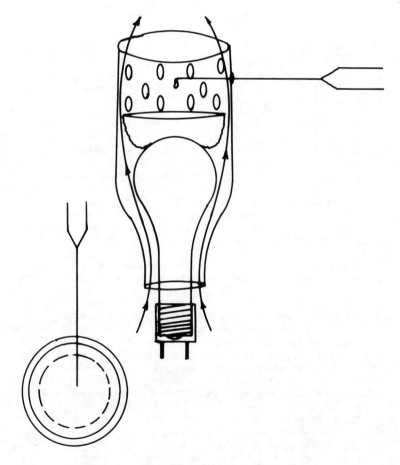

Figure 4. Formaldehyde generator oven.

Table IV.   Georgia-Pacific Formaldehyde Recovery Studies
@ 0.50 ACPH, 25$\pm$1°C, 50$\pm$4% RH

Target HCHO Conc. (ppm HCHO)	Actual Conc. (ppm HCHO)	% Recovery
0.40	0.37$\pm$0.03	92.5
0.30	0.28$\pm$0.02	93.3

A summary of Georgia Tech and G-P formaldehyde chamber
recoveries in Table V indicates good agreement between chambers, and
there appears to be no major formaldehyde losses within the chambers
considering experimental error of the techniques used.

Table V.   Summary Of Recovery Studies by Georgia Tech (GIT) and
Georgia-Pacific (G-P)

Lab	Target Chamber Conc. (ppm HCHO)	% RECOVERY
GIT	0.10	92.5
GIT	0.10	92.8
GIT	0.40	92.0
GIT	0.40	95.5
G-P	0.40	92.5
G-P	0.30	93.3

Board Conditioning For Large Chamber Testing

After analytical test methodology, board orientation within the
chamber, positive vs negative air displacement for make-up air to the
chamber, air make-up measurements, and environmental controls were
all evaluated and standardized, it became apparent to chamber
operators that board preconditioning was a very important factor in
obtaining comparable chamber results on identical board samples.
Tables VI-A & VI-B provide data on laboratories' A and B chamber
round-robin before and after proper conditioning facilities and
procedures were standardized. As can be seen in Table VI-A, the
relationship of chamber concentrations between Lab A and Lab B on
matched board sets before preconditioning procedures were established
varied between 25 to 67%. After preconditioning procedures were
established and carefully followed, the variation of chamber
concentrations between Lab A and Lab B dramatically improved over
five fold for matched board sets, i.e. 0 to 13.5% as shown in Table
VI-B.

Table VI-A. Chamber Concentration Consistency
Before Proper Conditioning

Board Set	Lab A (ppm HCHO)	Lab B (ppm HCHO)
1	0.27	0.45
2	0.37	0.50
3	0.49	0.63

TABLE VI-B. Chamber Concentration Consistency
After Proper Conditioning

Board Set	Lab A (ppm HCHO)	Lab B (ppm HCHO)
2	0.37	0.42
3	0.49	0.53
4	0.45	0.48
5	0.26	0.26
6	0.38	0.38

Section 2.2 of FTM-2 specifies a 7 day $\pm$ 3 hour, 24+3°C,and
50+4% RH conditioning period. During this interval much of the free
formaldehyde remaining from the manufacturing process is off-gassed.
In a publication to be released, Dr. George E. Myers of the Forest
Products Laboratory in Madison, Wisconsin, hypothesizes that" both
formaldehyde diffusion and reversible interactions with wood
hydroxyls (formation/hydrolysis of wood hemiformals) play important
roles in the ultimate release from UF boards of formaldehdye that is
liberated by hydroyzing resin, resin-wood, and formaldehyde-wood
states"(23). We believe the rate of formaldehyde released by the
hydrolysis of the UF binder is very low and contributes a minor
amount of formaldehyde released to the air. This amounts to less
than 0.07 ppm at the H.U.D. loading and air change rate for either
particleboard or paneling. Formaldehyde emitted in the early stages
after manufacture is predominately from physically absorbed
formaldehyde and low molecular weight resin/wood compounds formed
during the curing process. The "preconditioning" of boards
effectively reduces the contribution of these variable sources to
where the longer term hydrolytically susceptable compounds are the
prime sources of formaldehyde emissions. When this point is reached,
there is relatively little change in formaldehyde emissions with time
when temperature and humidity remain constant.

The concept of a "baseline" originated during early large scale
chamber testing when the test panels were loaded directly into the
chamber with-out a conditioning period. The HCHO levels were
monitored over a period of several days. During that interval, it
was observed that there was a rapid decrease in HCHO levels over the
first few days, followed by a interval of relatively slow decrease.
This later interval usually exhibited a rate of formaldehyde
decrease of 2 to 3% per day. At this point panels were said to be at
"baseline" or steady-state formaldehyde equilibrium. Essentially,

the board panels were undergoing conditioning in the large test chamber. Environmental conditions within the chamber were a uniform 24°C, 50% RH, 0.50 air changes per hour, and the make-up air had a formaldehyde level of less than 0.02 ppm. Therefore, replicate sets of panels gave very similar results in different large chambers.

Figure 5A represents our early attempt to condition panels using an "open system". Seven cabinets 0.61 meters wide X 1.52 meters high X 3.05 meters long were placed 30.48 centimeters above the floor. An exhaust blower (Figure 5A,Item B) at the end of each cabinet pulls air through a flow equalizing baffle. Each blower discharges about 30 cfm of air to outside the building. The total flow from the seven cabinets resulted in a total air exchange in the building every 20 minutes. A large blower (Figure 5A,Item B) completely cycled any air not exhausted through the cabinets every 2 minutes. A diffuser grill (Figure 5A,Item G) spreads the air evenly across the room. This conditioning system appeared to condition test boards with results similar to those achieved by leaving them in the large chamber for seven days. Typically, boards showed a formaldehyde decrease of 2 to 3% per day after being loaded into the large chamber from the conditioning system.

Figure 5B represents our current design - the "closed system". This system is totally closed with all air filtered through PURAFILL II Chemisorbant. Only one blower (Figure 5B,Item B) is used to circulate the air. The purified air is discharged through a diffuser grill (Figure 5B,Item G). All the air passes through the cabinets and filtration system every two minutes. The air velocity across the panels averages 9.1 meters per minute which is about 10 times that of the "open system". These velocities are consistent with ASHRAE standards for satisfactory operations (24). A flow equalizing baffle assures even flow through all areas of the cabinet. The cabinets are also larger, 0.61 meters wide X 2.5 meters high X 3.05 meters long. The formaldehdye concentration in the air before passing through the cabinets ranges between 0.02 to 0.05 ppm formaldehyde depending on product mix of the test panels. The exit air is generally 20 to 60% higher in formaldehyde content than the purified make-up air depending on the initial emission level of the conditioning panels. If a set of boards is expected to be a high emitter, the boards can be positioned with its shortest axis across the air flow to minimize formaldehyde buildup in the air stream.

We have observed test panels conditioned in the "closed system", with the exception of high density and high emitting products, achieve a "baseline" at the end of the 7 days conditioning period.

Temperature Effect On Chamber Concentrations

In the FTM-2 "Formaldehyde Test Method for Large Scale Test Chamber", the method allows a temperature correction factor to be applied to formaldehyde concentrations determined at temperatures other than the desired 25+0.5°C. In addition, the states of Wisconsin and Minnesota allow temperature corrections of formaldehyde levels determined at temperatures other than 25°C for field complaint investigations. The temperature correction factors are based on the popular Berge' Equation (25).

To verify the Berge' temperature correction, an experiment at different chamber temperatures was performed on various types of wood

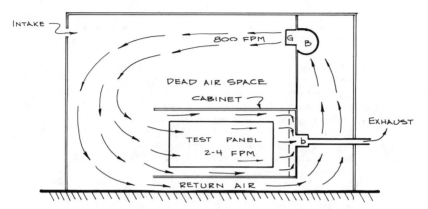

Figure 5A. "Open system" conditioning cabinet.

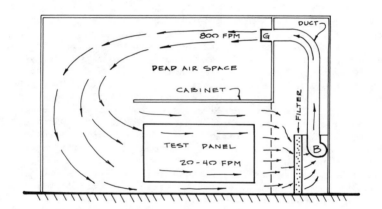

Figure 5B. "Closed system" conditioning cabinet.

products (i.e. particleboard, medium density fiberboard, paneling). Relative Humidity was controlled at a steady 50+4% for all temperatures. High formaldehyde concentrations to lower concentrations ratios were calculated for each product type at each corresponding temperature. The differences in the corresponding temperatures were plotted against the concentration ratio. The data obtained during this study is graphically represented in Figure 6. The derived relationship can be mathematically summarized in Equation 2.

$$Cn/Co = 0.7939 + 0.2358 \ \Delta T \qquad (2)$$

where Cn is new concentration at Tn
Co is initial concentration at To
Tn > To $\Delta T = Tn - To$
Tn = Higher temperature in Centigrade
To = Lower temperature in Centigrade

As can be seen in Figure 6, the correlation of the formaldehyde concentration ratio to a temperature difference is directly proportional. A statistical treatment of the data using a least-squares regression indicates a good correlation with a coefficient of 0.91.

Table VII presents a comparison of the experimentally derived temperature correction factor to the Berge' factor. The calculated Berge' factor is based on a temperature coefficient of 9799 recommended in the FTM-2 method. Based on this limited data base, it appears the temperature correction for formaldehyde concentrations is independent of product type, and the Berge' calculated factor appears to be about 7-10% too low for a temperature difference greater than 2°C.

Table VII. Temperature Correction for Formaldehyde
Chamber Concentrations

Temperature Difference(°C)	Derived Correction From Figure 8	Berge' Correction Calculated
1deg.	1.03	1.11
2deg.	1.27	1.25
3deg.	1.50	1.40
4deg.	1.73	1.57
5deg.	1.97	1.76

Effects of Loading And Air Exhchange Rate on Chamber Concentration

In this section of the paper, data on the effects of loading and air exchange rate on formaldehyde concentration in large scale test chambers will be presented. This data has been obtained on UF bonded wood products such as particleboard, medium density fiberboard, and hardwood plywood paneling from several laboratory test chambers. The purpose of presenting this data is to give you a general idea of the impact of product loading and air exchange rate on chamber concentrations. Dr. George M. Myers in a recent publication(26) discusses this subject in mathematical terms based on J.J Hoetjer's theoretical model for formaldehyde emissions from composition board (27). This article presents an in-depth theoretical discussion of this topic that cannot be covered in this paper. Some of the data used by Dr. Myers in his article comes from this same data base.

Table VIII presents chamber data on underlayment particleboard, mobile decking particleboard, and industrial particleboard obtained from four different chambers identified A, B, C and D . A particleboard "set" is a specific production run of a particleboard type. The observed concentration is the formaldehyde level actually determined in the chamber for a specific loading and air change rate. "N" represents the air change rate (number per hour). The column labeled "L" is the loading (m2/m3) that the test was conducted. The column "N/L" (m/hr) is the ratio of air change rate to the loading. Finally, the column labeled "Normalized Chamber Concentration" is the actual chamber concentration (first column) normalized to 0.3 ppm at N/L = 1.16. The 0.3 ppm chamber concentration at 0.43 m2/m3 loading and 0.5 air changes per hour is the H.U.D. formaldehyde standard for particleboard. Figure 7 graphically represents the normalized formaldehyde chamber concentrations to loading at air changes of 0.5, 1.0 and 1.5. The points which define the curves are averages of the normalized concentrations.

Table IX presents chamber data obtained in only one large test chamber identified as A on medium density fiberboard made at one plant. A medium density fiberboard "set" is a specific production run. The columns are labeled the same as the particleboard Table VIII described above. The "Normalized Chamber Concentration" is based on a 0.6 ppm formaldehyde concentration at an N/L ratio of 0.96. The choice of 0.6 ppm concentration is purely arbitrary. Figure 8 graphically represents the normalized formaldehyde chamber concentrations to loadings at air changes of 0.5, 1.0 and 1.5. The points which define the curves are averages of the normalized concentrations.

Figure 9 presents chamber data of only one set of hardwood plywood paneling performed at different loading and air change rates.

As can be seen in Figures 7, 8 and 9, the air change rate influences ambient formaldehyde levels more than does loading above 0.2 m2/m3 for any of the wood products. However, for loadings below 0.2 m2/m3, the major influence on formaldehyde levels is loading. In addition, the effect of ventilation rate on chamber concentration is different for each wood product type, i.e. particleboard, medium density fiberboard, hardwood plywood paneling.

These curves provide an important clue to the effect of lowering air change rate and increasing the amount of emitters in energy efficient dwellings. Over the past 11 years, fresh air changes have steadily declined to save energy until in some instances the air changes are below 0.15 m3/minute per occupant recommended by ASHRAE for health in an office environment where smoking is not permitted (28). It is obvious that continuous decreasing of air infiltration will continue to increase indoor air pollution from sources which are potentially alot worse than formaldehyde, i.e. insecticides, cleaners, oxides of nitrogen, carbon monoxide, biological contaminants, etc. A majority of our

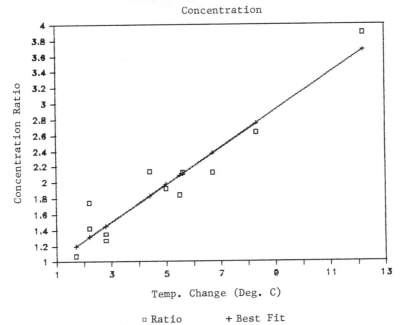

Figure 6. Temperature effect on chamber formaldehyde concentration.

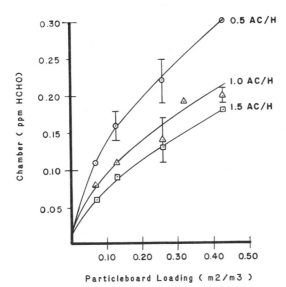

Figure 7. Effect of air change rate and loading on chamber formaldehyde concentration - particleboard.

Table VIII. Particleboard - Loading and Air Exchange Rate Effects
on Chamber Concentration

	Observed Conc. (ppm)	N (AC/H)	L (m2/m3)	N/L (m/Hr.)	Normalized By Calculation Conc.(ppm HCHO)
Set 1	0.26	0.5	0.26	1.92	0.22
Lab A	0.13	1.0	0.26	3.85	0.11
Set 2	0.14	1.0	0.26	3.85	0.14
Lab A	0.22	0.5	0.26	1.92	0.22
Set 3	0.18	0.5	0.26	1.92	0.22
Lab A	0.11	1.0	0.26	3.85	0.13
Set 4	0.23	0.5	0.26	1.92	0.22
Lab A	0.12	1.0	0.26	3.85	0.11
Set 5	0.21	0.5	0.43	1.16	0.30
Lab A	0.13	0.5	0.32	1.56	0.19
Set 6	0.15	0.5	0.16	3.12	0.14
Lab A	0.31	0.5	0.43	1.16	0.30
Set 7	0.26	1.0	0.43	2.33	0.22
Lab B	0.18	1.0	0.26	3.85	0.15
	0.13	1.0	0.13	7.69	0.11
	0.09	1.0	0.07	14.29	0.08
	0.35	0.5	0.43	1.16	0.30
	0.26	0.5	0.26	1.92	0.22
	0.17	0.5	0.13	3.85	0.15
	0.13	0.5	0.07	7.14	0.11
	0.21	1.5	0.43	3.48	0.18
	0.15	1.5	0.26	5.77	0.13
	0.10	1.5	0.13	11.54	0.09
	0.07	1.5	0.07	21.43	0.06
Set 8	0.43	0.50	0.43	1.16	0.30
Lab C	0.32	0.50	0.26	1.92	0.22
	0.23	0.50	0.13	3.85	0.16
	0.28	1.00	0.43	2.33	0.20
	0.22	1.00	0.26	3.85	0.15
	0.16	1.00	0.13	7.69	0.11
Set 9	0.12	1.00	0.43	2.33	0.19
Lab D	0.09	1.00	0.26	3.85	0.14
Set A	0.12	0.50	0.26	1.92	0.19
	0.19	0.50	0.43	1.16	0.30
Set B	0.07	1.00	0.43	2.33	0.19
	0.05	1.00	0.26	3.85	0.14
	0.06	0.50	0.26	1.92	0.16
	0.11	0.50	0.43	1.16	0.30
Set C	0.23	1.00	0.43	2.33	0.20
	0.19	1.00	0.26	3.85	0.17
	0.31	0.50	0.26	1.92	0.27
	0.34	0.50	0.43	1.16	0.30
Set D	0.10	1.00	0.26	3.85	0.20
	0.07	1.00	0.13	7.69	0.14
	0.09	0.50	0.13	3.85	0.18
	0.15	0.50	0.43	1.16	0.30

Table IX. Medium Density Fiberboard - Loading and Air Exchange Rate
Effects on Chamber Concentration

	Observed Conc. (ppm)	N AC/H)	L (m2/m3)	N/L (m/Hr.)	Normalized By Calculation Conc.(ppm HCHO)
Set 1	0.66	0.50	0.52	0.96	0.60
Lab A	0.47	1.00	0.52	1.92	0.43
	0.30	1.00	0.26	3.84	0.27
	0.53	0.50	0.26	1.92	0.48
	0.29	0.50	0.13	3.84	0.26
	0.16	1.00	0.13	7.69	0.15
Set 2	0.84	0.50	0.52	0.96	0.60
Lab A	0.51	1.00	0.52	1.92	0.36
	0.33	1.00	0.26	3.84	0.24
	0.50	0.50	0.26	1.92	0.36
	0.21	1.00	0.13	7.69	0.15
	0.41	0.50	0.13	3.84	0.29
Set 3	0.83	0.50	0.52	0.96	0.60
Lab A	0.76	1.00	0.52	1.92	0.55
	0.34	1.00	0.26	3.84	0.24
	0.43	0.50	0.13	3.84	0.31
	0.27	1.00	0.13	7.69	0.20
Set 4	1.00	0.50	0.52	0.96	0.60
Lab A	0.76	1.00	0.52	1.92	0.46
	0.37	1.00	0.26	3.84	0.22
	0.51	0.50	0.26	1.92	0.31
	0.37	0.50	0.13	3.84	0.22
	0.26	1.00	0.13	7.69	0.16
Set 5	0.45	0.50	0.43	1.16	0.50
Lab A	0.46	0.50	0.43	1.16	0.50
	0.34	1.00	0.43	2.32	0.38
	0.16	1.00	0.13	7.69	0.18
	0.27	0.50	0.13	3.84	0.29
	0.13	1.00	0.13	7.69	0.14
Set 6	0.85	1.00	0.43	2.32	0.34
Lab A	0.87	1.00	0.43	2.32	0.34
	1.25	0.50	0.43	1.16	0.50
	0.37	1.00	0.13	7.69	0.15
	0.54	0.50	0.13	3.84	0.22
Set 7	0.69	1.00	0.43	2.32	0.45
Lab A	0.77	0.50	0.43	1.16	0.50
	0.28	1.00	0.13	7.69	0.18
	0.39	0.50	0.13	3.84	0.25
Set 8	0.37	1.00	0.43	2.32	0.40
Lab A	0.46	0.50	0.43	1.16	0.50
	0.16	1.00	0.13	7.69	0.18
	0.24	0.50	0.13	3.84	0.26
Set 9	0.22	0.50	0.13	3.84	0.19
Lab A	0.16	1.00	0.13	7.69	0.14
	0.57	0.50	0.43	1.16	0.50
	0.45	1.00	0.43	2.32	0.40
	0.26	0.50	0.13	3.84	0.23

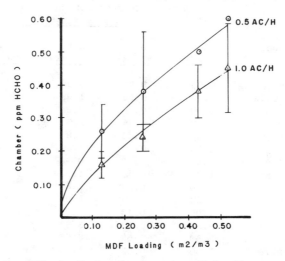

Figure 8. Effect of air change rate and loading on chamber formaldehyde concentration - medium density fiberboard.

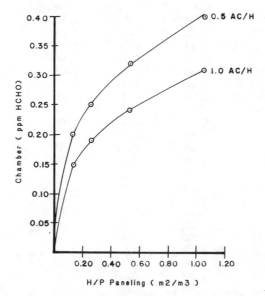

Figure 9. Effect of air change rate and loading on chamber formaldehyde concentration - hardwood plywood paneling.

field investigations points to the major problem - "a lack of fresh air infiltration into the living space". Since formaldehyde is so easy to analyze and is so ubiquitous, many field investigators have assigned any non-specific acute symptoms to formaldehyde regardless of the level of formaldehyde exposure. This simplistic approach is both dangerous and naive on the part of the investigator. The total indoor environment must be evaluated before any causation can be even speculated.

Combination Loading Of Different Wood Products

Two Product Loading - Particleboard And Hardwood Plywood Paneling. The effects of mixing particleboard and hardwood plywood paneling on chamber concentration at a particular product loading and a given air change rate is of practical importance. It is seldom that a single formaldehyde emitting product is ever used alone in a dwelling. Therefore, it would be desirable to predict the final chamber concentration when more than one formaldehyde emitting product is combined.

The H.U.D. formaldehyde standards of 0.2 ppm and 0.3 ppm for hardwood plywood paneling and particleboard, respectively, were chosen because the combination of these products at their specific loadings and air change rate would result in a chamber concentration of less than 0.4 ppm. This assumption was based on four studies. The first was the Clayton Study (29) sponsored by H.U.D. in which four mobile home units were constructed with wood products of known formaldehyde emission characteristics as determined in the large scale chamber. The other three studies were from an association and two industrial laboratory chambers working independently of each other. Essentially, all four studies came to the same conclusion - it is possible to predict chamber concentrations from a combination of two formaldehyde emitting products.

Mr. William Groah of the Hardwood Plywood Manufacturers Association suggested an empirical method to predict the chamber concentration of a two wood product combination. He suggested plotting the observed chamber combination against the arithmetic total of the individual chamber concentration. Figure 10 graphically represents the two product mix based on the data in Table X. As can be seen in Figure 12, there is a very good correlation (R^2 = 0.98) using this approach.

Three Product Loading - Particleboard, Hardwood Plywood Paneling and Unfinished Medium Density Fiberboard

An investigation of a three product combination was conducted in the Decatur Chamber. A linear relationship with a correlation cofficient (R^2) of 0.99 indicated the empirical relationship established for a two product combination also holds for a three wood product combination. Figure 11 presents a graphical summary of the observed data in Table XI.

Table X. Two Product Loading Chamber Formaldehyde Levels

Particleboard* Alone (ppm HCHO)	Paneling** Alone (ppm HCHO)	Arithmetic Total (ppm HCHO)	Observed Combined Product Conc. (ppm HCHO)
0.19	0.70	0.89	0.69
0.32	0.54	0.86	0.66
0.23	0.31	0.54	0.36
0.19	0.13	0.32	0.20
0.08	0.29	0.37	0.29
0.19	0.19	0.38	0.24
0.23	0.58	0.81	0.59
0.75	0.20	0.95	0.70
0.28	0.08	0.36	0.23
0.40	0.40	0.80	0.60
0.40	0.15	0.55	0.41
0.31	0.20	0.51	0.33
0.53	0.29	0.82	0.50

Note: * Particleboard Loading = 0.43 m2/m3
     ** Paneling Loading   = 0.95 m2/m3
        ACPH   = 0.5
        Temperature   = 25+1°C
        R.H.   = 50+4%

Table XIV. Three Product Loading Chamber Combination

Particleboard* Alone (ppm HCHO)	H/P Paneling** Alone (ppm HCHO)	MDF*** Alone (ppm HCHO)	Arithmetic Total (ppm HCHO)	Observed Combined Product Conc. (ppm HCHO)
0.23	0.58	0.16	0.97	0.54
0.23	0.58	0.34	1.15	0.65
0.19	0.19	0.29	0.67	0.40

Note: *   Particleboard Loading   = 0.43 m2/m3
     ** Paneling Loading   = 0.95 m2/m3
     *** Medium Density Fiberboard   = 0.43 m2/m3
         Temperature   = 25+1°C
         ACPH   = 0.5
         Relative Humidity   = 50+4%

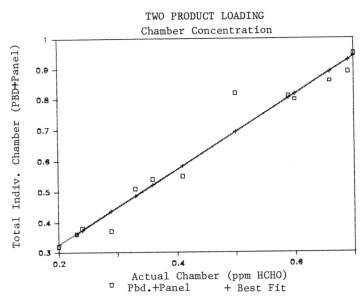

Figure 10. Effect of two dissimilar wood products on chamber formaldehyde concentration.

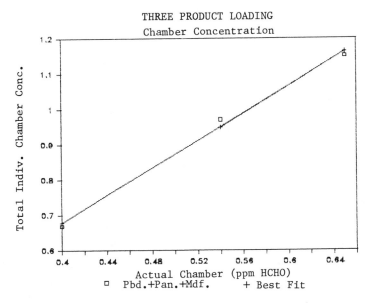

Figure 11. Effect of three dissimilar wood products on chamber formaldehyde concentration.

Round Robin Chamber Comparisons

During the past three years, Georgia-Pacific laboratories have
participated in thirty-three inter-large scale test chamber round
robin tests. As stated in the Conditioning Section, early round
robins showed poor correlation between chambers. However, when the
large scale test chamber methodology became standardized with the
issue of FTM-2, the relationship between chambers steadily improved.
Now, improved board conditioning procedures, attention to analytical
technique, and standardized chamber construction have improved the
relationship between chambers.

Figure 12 is a graphical representation of thirty-three
individual chamber round robins between Georgia-Pacific's chambers in
Decatur, Georgia and Sacramento, California to various test chambers
identified as A, B, C, D. The data obtained in this three year study
were based on both an exchange of the same boards or statistically
sampled matched board sets. This data includes 10 paneling sets, 15
particleboard sets, 8 medium density fiberboard sets. There were
four tests that involved testing the very same boards.

The relationship of the Georgia-Pacific chambers to the other
four chambers in this study indicates good agreement with a
coefficient of correlation of 0.94. The major conclusion from this
study is that chamber tests are reproducible provided the tests are
conducted under a strict test protocol.

Quality Control Test Methods and Chamber Correlations

The H.U.D. standard for U-F bonded particleboard and hardwood plywood
paneling requires in-plant monitoring of formaldehyde emissions from
these products with a quality control method that correlates to the
large scale test chamber. The most popular Q.C. test method used in
the U.S.A. is the Two Hour Desiccator Method, FTM-1. This method has
wide acceptance because of its simplicity and short test duration
required for in-plant monitoring. G-P uses the 2 Hour Desiccator
Method for in-plant monitoring of its hardwood plywood finished
paneling (e.g. print, paper overlay, and veneer). Even though we
use the 2 Hour Desiccator for particleboard, our particleboard plants
have had success over the past seven years with a method known as the
"Equilibrium Jar Method" published internally as GPAM 203.6. Other
methods Georgia-Pacific have evaluated are the Formaldehyde Surface
Emission Monitor (FSEM) and the small scale test chamber (SSTC)
developed by Oak Ridge National Laboratories under a project funded
by the Consumer Products Safety Commission.

Our experience in developing correlations for quality control
test values to matching chamber concentrations has shown that each
correlation varies for each product type and wood manufacturing
unit. When we speak of product type in the U.S. particleboard
industry, we are referring to particleboard that is classified by its
end use (e.g. floor underlayment particleboard; mobile home decking
particleboard; industrial particleboard). Even though these
different types of particleboard may be made on the same equipment
and with the same binder system, it is the desired physical
properties and manufacturing variables that can influence the
emission characteristic of the board. Some researchers call this
emission characteristic emissitivity or interphase transport

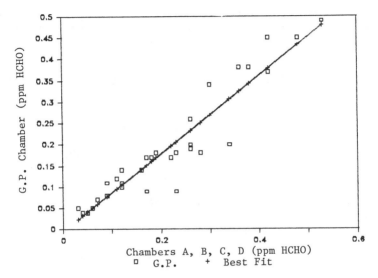

Figure 12. Chamber round robin tests.

parameter. In short, it is all related to how the board is
constructed. The important thing to remember is that correlations
must be developed for each product. They may or may not relate to
other manufacturing units making the same product.

General Methodology Used To Correlate A Q.C. Method to the Large Scale Chamber

Upon completion of the chamber test, the hardwood plywood paneling or
particleboard is removed and 12 each 7.00cm x 12.7 cm specimens are
randomly cut from each board loaded into the chamber. For the
surface monitor (FSEM) and the small scale test chamber(SSTC), one
30.5cm x 30.5cm board is cut from each board loaded in the chamber.
These samples are immediately tested by the Equilibrium Jar for
particleboard or the Two Hour Desiccator or FSEM or SSTC for all wood
product types. The values obtained from each test are averaged and
are then compared to the chamber concentration observed for that
loading and air change rate.

   At Georgia-Pacific, no conditioning period is observed for any
board specimen after it is removed from the chamber. The purpose of
this procedure is to determine the precise emission characteristic of
the board at the time of the chamber formaldehyde determination. The
FTM-1 and FTM-2 procedures dictate that the small specimens are cut
and conditioned along with the large boards prior to the chamber
test. It has been observed by G-P that conditioning small specimens
gives different 2 Hour Desiccator or Equilibrium Jar values from
those obtained from specimens cut from whole boards and panels
conditioned in a similar manner. Based on our experience, this
difference is not as large for low emittng particleboard as it is for
freshly finished paneling.

   Moisture content of each specimen is determined and recorded
after completion of the secondary tests. In the case of
particleboard, the moisture content ranges between 7 to 9% by
weight. Paneling moisture content usually ranges between 8 to 10% by
weight. The moisture pick-up in the wood specimens tested by the 2
Hour Desiccator generally runs less than 0.2% by weight.

   All quality control tests and specimen conditioning are
conducted under carefully controlled environmental conditions, i.e.
temperature = $24 \pm 0.5°C$, $50 \pm 5\%$ relative humidity and a background
formaldehyde level of less than 0.1 ppm. Ourselves as well as others
have found that temperature effects on the quality control test
values follow the same pattern observed in the large scale chamber
(30). In short, the Berge' temperature correction can be applied to
the quality control test methods.

Equilibrium Jar Method (GPAM 203.6)

   The Equilibrium Jar method is based upon the collection of
formaldehyde in an empty 1 liter jar placed mouth to mouth on top of
the second jar containing one particleboard sample 7.00cm x 12.70cm
with all edges wax sealed. The loading ratio in this method is 13.3
m2/m3. At the end of a 24 hour eqilibration time, the two jars are
separated and the formaldehyde in the top jar is swept into a 0.1 N
sodium hydroxide absorbing solution. The collected formaldehyde is
then analyzed using the chromotropic acid procedure described in
NIOSH P&CAM 125. Results are expressed in ppm (vol./vol.).

Thirty-one particleboard sets of type 1 were obtained from all of our particleboard plants. As you can see in Figure 13, there is a cloud of points below 0.3 ppm chamber and a group of three points around 0.4 ppm. The cloud of points around 0.3 ppm represents current production which is made to meet the H.U.D. particleboard standard. The three points around 0.4 ppm are from a special plant test performed to define the shape of the Equilibrium Jar/chamber correlation curve. The reasons we can plot all the data points from all the plants are: 1) all plants have the same process; 2) we have a historical data base.

The correlation of the Equilibrium Jar to the chamber has historically been a good fit. In the case for type 1 particleboard, the relationship is a linear one with a good correlation coefficient (r^2) of 0.86.

Current inter- and intra-laboratory evaluations indicate the Equilibrium Jar's precision is +8% and between laboratory variation is about +10%.

Two Hour Desiccator Method, FTM-1

Specimens of particleboard or paneling are placed on a plate in a 10 liter desiccator containing an inverted 300 ml beaker with a petri dish top containing 25 ml. of water. The number of 7.00cm x 12.70cm waxed edged specimens placed in the desiccator is eight. The samples remain in the closed desiccator for exactly 2 hours. At the end of that time, the desiccator is opened and the 25 ml. of water is analyzed for formaldehyde using the chromotropic acid procedure described in P&CAM 125. The solution is analyzed in triplicate and the average value in micrograms of formaldehyde per milliliter (ug/ml) is reported.

Figure 14 provides a graphical representation of 27 chamber tests conducted on a variety of veneer, print, a paper overlay finished hardwood plywood paneling. Even though it is not shown, a breakdown by different product type did not affect the correlation by anymore than 5%. As with particleboard, the cloud of points below 0.2 ppm represents current production made to meet the H.U.D. hardwood plywood paneling standard (0.2 ppm chamber). The group of points between 0.24 and 0.36 ppm chamber are from earlier chamber studies needed to define the curve.

A linear regression using the least-squares method gave an good 0.86 correlation coefficient (r^2). As can be seen in Figure 14, there appears to be more scatter in the data than in the particleboard graph. This may be due to the heterogeneous nature of plywood.

The precision of this method on the same samples appears to be within +6%. The variation between laboratories is about +10%.

Formaldehyde Surface Emission Monitor (31)

Even though we did several tests evaluating the FSEM to the chamber, the data we obtained was not convincing enough to continue work on this methodology. Work performed at Georgia Institute of Technology, Georgia Tech Research Institute, reflected the same problems and observations we experienced using this methodology. A summary of Georgia Tech's conclusions on this methodology (32) is as follows:

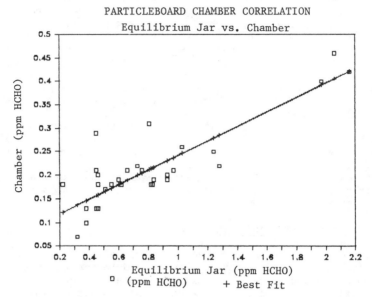

Figure 13. Correlation of equilibrium jar to chamber for particleboard.

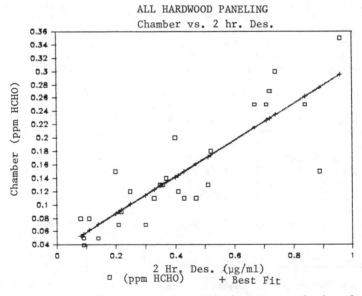

Figure 14. Correlation of 2 hour desiccator to chamber for hardwood plywood paneling.

1. The accuracy and precision of the chemistry associated with this anaylsis is extremely good when experimental variables are well controlled;

2. Interboard variations of FESM measurements on particleboard averaged 30.5%;

3. An intraboard variation of FESM measurements on particleboard was 22%;

4. Since the accuracy and precision data of the chemical analysis process are very good, these FESM variations are a function of wood product characteristics and/or errors originating in the FESM methodology employing molecular sieve 13X;

5. The significant interboard and intraboard variations in FESM measurements indicate that this technique cannot be accurately used to measure and distinguish between wood product formaldehyde emission rates; and

6. If a statistically large number of FESM measurements are made per wood product and if the formaldehyde emission rate characteristics are significantly different (high versus low), the FESM data might qualitatively distinguish between them.

Small—Scale Test Chamber

Our Small Scale Test Chamber (SSTC) is constructed of 2 cm thick plywood. The interior of the small chamber is lined with the same aluminum used in the large scale test chamber. The exterior surface is painted with an epoxy paint. Dimensions of our SSTC are similar to the SSTC developed at Oak Ridge National Laboratories. The interior dimensions are 63 cm x 63cm x 61cm with an internal volume of 241 liters. The internal equipment occupies about 2 liters thus giving an adjusted volume of 239 liters.

Temperature and humidity are controlled to $24\pm1°C$ and $50\pm4\%$. The fresh air entering the SSTC is filtered clean of all organic gases using PURAFIL II Chemisorbant. The amount of air entering the SSTC is controlled with a calibrated Brooks flowmeter which is equipped with a flow controller. A very small 7.5 cm diameter electric fan directed toward an air deflector provides the required mixing. The electric motor is totally enclosed. Air exiting the chamber is exhausted to the surrounding environment.

Samples are placed in the SSTC and the air flow adjusted to give an N/L ratio of 2.19 m/hr. An ambient air sample is obtained at four hours after loading the board into the SSTC. Another air sample is pulled 24 hours after the boards were loaded. If the two concentrations agree, this value is reported as the SSTC concentration for that product. The method used to determine the formaldehyde is the chromotropic acid procedure as described in NHIOSH P&CAM 125 except only 0.5 liters per minute for 45 minutes is used for the flowrate for air sampling.

Of the new methodologies being studied, the Small Scale Test Chamber seems to hold the most promise. The obvious advantage of this methodology is it more closely approximates the large scale test chamber in operational characteristics-interaction of board and air. In the 2 Hour Desiccator method, the presence of water in an enclosure is a compounding factor that is not fully understood. However, our experience with SSTC on all types of particleboard and medium density fiberboard indicates good correlation at N/L = 1.16 and 2.19 m/hr to the large scale test chamber. At this time we have only looked at these two N/L values. Paneling on the other hand has not shown a clear correlation to the large scale chamber at a N/L of 2.19 m/hr. Interestingly, this is the same observation that Georgia Tech has seen in their study of SSTC vs Large Test Chamber. It appears that the small scale chamber, like other small test methods, is influenced by product homogeneity and perhaps a scale down factor. At this point in time, formaldehyde emission rates determined by the SSTC should be cautiously used in predicting ambient formaldehyde concentrations.

Quality Control Methods Conclusion

Based on our experience, it appears that a quality control method which correlates to the chamber for a particular product type does not always work for all products. The only universal test method for all products is the large scale test chamber. A quick and reliable formaldehyde quality control test method is becoming more important as formaldehyde levels in the chamber fall below 0.15. A universal small scale test method (Q.C.) does not seem to exist at this time. However, the Small Scale Test Chamber may be the closest to fulfilling that purpose.

Actual Chamber Concentration Vs Quality Control Predicted Concentration

This study evaluated the effectiveness of how well the correlation of a quality control method to the chamber predicted actual formaldehyde chamber concentrations from freshly manufactured board.

Particleboard and paneling samples were pulled from the manufacturing line shortly after it was made or finished. A portion of the boards was analyzed by the plant Q.C. laboratory personnel without being told the purpose of the test. The boards were transported to the Decatur laboratory within 24 hours after manufacture. The boards were conditioned for 24 hours upon arrival at the laboratory, and the following day they were inserted into the chamber.

The ambient formaldehdye concentration was determined within the chamber another 24 hours later. In the meantime, the Q.C. laboratory was called and their Equilibrium Jar or 2 Hour Desiccator value for that board was used to determine the corresponding chamber concentration from their correlation. A total of 6 test sets were evaluated in this manner and the results are summarized in Table XII. It is clear the predicted chamber concentration for fresh board relates well with actual chamber concentration.

Table XII. Fresh Board Study Actual Chamber Concentration Vs
            Quality Control Predicted Chamber Concentration

Test	Actual Chamber (ppm HCHO)	Q.C. Predicted Concentration (ppm HCHO)
1	0.42	0.40
2	0.45	0.43
3	0.28	0.23
4	0.13	0.12
5	0.17	0.13
6	0.10	0.13

Field Measurements Vs Predicted Formaldehyde Levels

Actual formaldehyde measurements made while performing field
investigations using the CEA 555 Air Monitor were corrected to
25°C. Wood samples removed from the investigation site were
returned to the laboratory, and the corresponding quality control
test method was used to determine formaldehyde content of the
specific wood product. The formaldehyde value obtained from the
quality control test method was then used to determine the chamber
concentration from the established correlations (Figures 13 & 14).
As can be seen in Figure 15, the linear regression using the least
squares methods on the eighteen field tests, there is a definite
relationship of field measurements to predicted chamber
concentrations based on quality control tests performed on samples
obtained in the field. This relationship is more than coincidence
because it indicates to us that our correlations can predict ambient
formaldehyde levels in the real world once the various emitting
substances are identified. However, this ability to identify the
emitting substances takes product knowledge, training and
experience.

Conclusions

The following conclusions can be drawn from this work:

1. Formaldehyde concentrations observed in an environmental
   chamber do relate to real world formaldehyde
   levels provided conditions are comparable.

2. The modified NIOSH Method P&CAM 125 and CEA 555
   accurately determine formaldehyde concentration found in
   living spaces.

3. Chamber formaldehyde recoveries are within analytical
   precision.

4. Strict adherence to conditioning procedures reduces
   between chamber variation.

5. Chamber concentrations of product combinations can be
   predicted empirically.

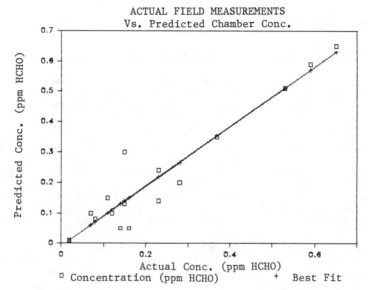

Figure 15. Actual field concentrations vs predicted chamber
concentrations from quality control test correlations.

6. Effects of loading and air exchange rates on chamber formaldehyde concentrations can be predicted.

7. Temperature effect on chamber concentrations can be predicted.

8. Equilibrium jar Q.C. test for a G-P particleboard type 1 correlates to large scale chamber.

9. 2 Hour Desiccator for G-P hardwood plywood paneling correlates to the large scale test chamber.

10. Product type may influence correlation of Q.C. test to large scale chamber.

11. Correlated Q.C. test can predict chamber concentrations regardless of board age.

12. Properly selected field specimens relate to actual field measurements.

<u>Literature Cited</u>

1. "Manufactured Home Construction and Safety Standards", U.S. Department of Housing and Urban Development, Federal Register, Vol. 49, No. 155, 8/9/84, 24CFR3280.
2. Sundin, B. "Present Status of Formaldehyde Problems and Regulations"; International Particleboard Symposium No. 16: Washington State University, Pullman, 1982.
3. Brummel, R.; "Oct. 1984. Report on European Trip". Personal correspondence, Jan. 16, 1985.
4. "Large Chamber Method - FTM-2-1983"; National Particleboard Association/Hardwood Plywood Manufacturers Association, 1983.
5. "Small Scale Test Method for Determining Formaldehyde Emission from Wood Products - Two Hour Desiccator Test - FTM-1-1983"; National Particleboard Association/Hardwood Plywood Manufacturers Association, 1983.
6. "Equilibrium Jar Method, Colorimetric Determination of Formaldehyde Using Chromotropic Acid Reagent for Product Testing - GPAM 203.6"; Georgia Pacific Corporation, 1979.
7. "Tentative Test Method for Emission of Formaldehyde from Wood Products - 24 Hour Desiccator Method"; National Particleboard Association, 1980.
8. Newton, L. "Formaldehyde Emissions from Wood Products: Correlating Environmental Chamber Levels to Secondary Laboratory Tests"; International Particleboard Symposium No. 16; Washington State University, Pullman, 1982.
9. Walker, J.F. "Formaldehyde"; 3rd Reinhold, London, 1964; p. 98.
10. Ibid, 1964; p. 98.
11. "Physiological Principles, Comfort, and Health"; ASHRAE Handbook Fundamentals, 1981; pp. 8.4-8.13.
12. Grot, D. "Plan for Testing Model for Formaldehyde Emissions from Pressed Wood Products"; National Bureau of Standards, July, 1984.

13. Lagus, P.L. "Air Leakage Measurements by the Tracer Dilution Method - A Review"; ASTM STP 719; Hunt, C.; King, J.; Trechsel, H.; Eds.; American Society for Testing and Materials, 1980; pp. 36-49.
14. "Large Formaldehyde Chamber Task Force Meeting-June 13, 1984"; Hardwood Plywood Manufacturers Association, Reston, Virginia.
15. Ibid., June, 1984.
16. "Air Chamber Test Method for Certification and Qualification of Formaldehyde Emission Levels"; U.S. Department of Housing and Urban Development, Federal Register, Vol. 49, No. 155, 8/9/84, 24CFR3280.406.
17. C.E.A. Instruments, Inc. "CEA 555 - Formaldehyde in Air"; FO-1 to FO-3; CEA Instruments, Inc., 15 Charles Street, Westwood, NY.
18. Lyles, G.; Dowling, F.; Blanchard, V.; "Quantitative Determination of Formaldehyde in the Parts Per Hundred Million Concentration Level"; J. Air Pollution Control Association, Vol. 15, No. 3, 1965; pp. 106-108.
19. Black, M.S. "Studies of Molecular Sieve 13X Solid Absorbent the Formaldehyde Surface Emission Monitor"; Georgia Institute of Technology, January 6, 1984.
20. Ibid., January 6, 1984.
21. Black, M.S. "Correlation Of Wood Product Formaldehyde Emission Rates As Determined Using A Large Scale Test Chamber, Small Scale Test Chamber, and Formaldehyde Surface Emission Monitor"; Georgia Institute of Technology, April 18, 1985.
22. Lehnman, W. "Formaldehyde Vapor Generation Using Syringe Pump Method"; Weyerhaeuser, Tacoma, Washington, 1983.
23. Myers, G.E. "Mechanisms of Formaldehyde from Bonded Wood Products" in this book.
24. "Standards for Satisfactory Conditions"; ASHRAE Handbook of Fundamentals, 1972, pp. 452-453.
25. Berge, A.; Mellegard, P.; Hanetho, O.; Ormstad, E." "Formaldehyde from Particleboard - Evaluation of a Mathematical Model"; Holz als Roh-und Werksoff, 1980, 38, pp. 251-255.
26. Myers, G. "Effect of Ventilation Rate and Board Loading on Formaldehyde Concentration: A Critical Review of the Literature"; Forest Products Journal, 1984, 34, pp. 59-68.
27. Hoetjer, J.J. "Introduction to a Theoretical Model for the Splitting of Formaldehyde from Composition Board"; Report from Methanol Chemie Nederland, June 8, 1980.
28. "Ventilation for Acceptable Indoor Air Quality"; ASHRAE 62-1981.
29. Singh; Walcott, J.; St. Pierre, C.; Ferrel, T.; Garrison, S.; Groah, W. "Evaluation of the Relationship Between Formaldehyde Emissions from Particleboard Mobile Home Decking and Hardwood Plywood Wall Paneling Determined by Product Test Methods and Formaldehyde Levels in Experimental Mobile Homes"; Clayton Environmental Consultants, Inc., Report, Prepared on Contract No. AC-5222, H.U.D., March, 1982.
30. Rybicky, J.; Horst, K.; Kambanis, S.; "Assessment of the 2 Hour Desiccator Test for Formaldehyde Release from Particleboard"; Forest Products Journal, Sept., 1983, 33, pp. 50-54.

31. Matthews, T.G.; Daffron, C.R.; Corey, M.D. "Formaldehyde
     Surface Emission Monitor - Protocol I: Pressed Wood Products";
     ORNL/TM-8656, Oak Ridge National Laboratory, June, 1983.
32. Black, M.S. "Correlation of Wood Product Formaldehyde Emission
     Rates As Determined Using A Large Scale Test Chamber, Small
     Scale Test Chamber, and Formaldehyde Surface Emission Monitor";
     Georgia Institute of Technology, April 18, 1985.

RECEIVED January 14, 1986

14

Predicting Real-Life Formaldehyde Release by Measurement in the Laboratory

M. Romeis

Centre Technique du Bois et de l'Ameublement, 10 avenue de Saint-Mande, 75012 Paris, France

The purpose of this study was to evaluate laboratory formaldehyde release test methods for predicting real-life formaldehyde air concentrations, human exposure levels, and health risk. Three test methods were investigated: the European perforator test, the gas analysis method at $60^{\circ}C$ and 3% RH, and the gas analysis method at $23^{\circ}C$ and 55% RH. Different types of particle-board bonded with urea-formaldehyde and urea-melamine-formaldehyde resins were tested. The results were used to rank boards as a function of test method, conditioning, short-term humidity, and temperature variations during storage. Additional experiments were conducted in small experimental houses at a Dutch research institute. Our conclusions are that relative ranking of products is influenced by the test method and by change in relative humidity. The relationship between test method and release in real-life situations is not clear. In fact, it seems impossible to use laboratory measurements to predict real-life product performance of board if the board is not fully in equilibrium with the atmosphere.

Formaldehyde emission from particleboard has been studied at our laboratory for over 15 years. We search for an answer to the following question: Given the fact that amino-resin bonded wood products have the ability to release formaldehyde into indoor air when they are in use, what simple and rapid analysis method can be used at the time of manufacture to predict formaldehyde release under use conditions as quantitatively as possible? Obviously, the chosen method needs to be applicable for all types of boards that are available on the market.

Background

The presence of formaldehyde is due to the necessity to provide for an excess of aldehyde, in order to get good resin curing. It is well

0097–6156/86/0316–0188$06.00/0
© 1986 American Chemical Society

established that the ratio of formaldehyde to total nitrogen
compounds is related to the emission tendency of the finished product
(1).
        This excess aldehyde may be present in different chemical states.
Various hypothesis exist about these:
        1. Formaldehyde is in a free state, and we may predict its
emission by means of well known physical laws.
        2. Formaldehyde is combined with wood and may be displaced by
other reagents, such as water. In this case, water addition will
cause aldehyde release.
        3. Formaldehyde is absorbed in the water absorbed in the wood
cell wall; it may be released when the water vaporizes from the
board.
        4. Formaldehyde is an integral chemical part of the cured
adhesive; it may be released by hydrolysis.
        Whatever the hypothesis, it always involves excess aldehyde. This
is to say that:
        1. The formaldehyde content diminishes with time.
        2. Analysis of the total excess aldehyde will give the maximum
quantity of formaldehyde a board may release during its life, and
        3. Analysis permits estimation of the rate of its release, and,
taking into account the maximum value found above, prediction of the
release rate of the board.

Experimental

Numerous previous studies have led to equations permitting
predictions of formaldehyde release rates, but none of these were
based on boards manufactured in France, particularly not melamine-
urea-formaldehyde adhesive bonded boards. It was interesting to us
to apply these testing methods to French boards. We selected
industrial panels for this study, so that the results have practical
value. Unfortunately, this choice presents drawbacks in that in the
comparison of industrial panels several parameters may vary from one
panel to another.
        From the numerous possible methods available, three were
selected, because they had already given good correlations in other
European studies. These methods are:
        1. The perforator method, European Standard CEN EN 120 (2): This
method uses cubic specimens, 2 x 2 cm x board thickness. This is a
toluene total extraction method and the formaldehyde is determined by
titration with iodine. The result is expressed in HCHO mg/100 g
board.
        Two gas flow methods: These methods apply to larger specimens,
up to 9 x 50 cm x board thickness for our apparatus. Board edges may
be sealed. The aldehyde is driven off by nitrogen flow, recovered in
water, and determined photometrically with chromotropic acid. The
chosen methods are:
        2. The FESYP Gas flow method, using nitrogen at $60^{\circ}C$, 3%
relative humidity, 120 L/hour nitrogen. The result is expressed in
HCHO ug/kg board x hour (3), and
        3. The European Draft standard (4), using nitrogen at $23^{\circ}C$, 55%
RH, 20 to 60 L/hour nitrogen. The result is expressed in HCHO
ug/nitrogen liter.

We used eight boards, see Table I. All boards were 19 mm thick.
Each method will provide as a result a formaldehyde quantity. Thus,
it is possible to rank the boards in order of increasing values,
presumably corresponding to increasing "pollution".

Table I. Board Samples Used in this Study

Type	Wood Species	Adhesive	Year of Manufacture
F1	Mixed Hardwoods	UF #1; low F/NH_2 + scavenger	1983
F2		UF #1; without scavenger	1981 & 1983
F3		UF #2; high F/NH_2	1981
F4		Melamine-UF; high F/NH_2	1981
R1	Mixed Softwoods	UF #3; low F/NH_2	1984
R2		UF #4; equivalent to #1	1984
R3		UF #5; F/NH_2 between #4 & #2	1984
R4		Melamine-UF^2	1984

Results and Discussion

It is first necessary to check if the relative ranking of the samples
is the same for all analytical methods. If this is not the case,
then it will be necessary to take into account the emission rate. If
this does not explain possible discrepancies, it becomes necessary to
consider the influence of storage or conditioning, i.e. the history
of the board from the time of manufacturing in the press and the
influence of sudden changes of environmental conditions. The results
obtained with the three analyticlal methods are shown in Table II.

Table II. Results Obtained by Three Analytical Methods

Board		Perforator mg/100g board	Gas flow FESYP 60°C; 3% RH HCHO mg/kg board hr	Gas flow CEN 23°C; 55% RH HCHO mg/L nitrogen
F1		16	-	1.02
F2	1981	28	8.9	-
	1983	28		1.83
F3		73	3.8	-
F4	1981	64	10.9	-
	1983	61	-	1.33
R1		10	-	1.88
R2		16	-	1.83
R3		21	-	2.87
R4		38	-	1.08

The absence of any correlation between the three methods can be seen immediately. In particular, if we compare the results obtained with the perforator and the gas flow method at $23^{\circ}C$, 55% RH which is close to normal use conditions, we note the inconsistent results shown in Figure 1. In fact, if we calculate the ratio, perforator rate to gas flow rate, we obtain the following approximate ratios:

Boards R1, R2, and R3:   Ratio of Perforator to Gas Flow = 7
Boards F1 and F2:   Ratio of Perforator to Gas Flow = 15
Boards F4, and R4:   Ratio of Perforator to Gas Flow = 40

This raises doubts about the reliability of predicting formaldehyde emission by using the perforator. However, on the other hand, each group of products corresponds to a given adhesive. This means that for a given adhesive a constant relationship exists between the perforator rate and the emission, as has been already demonstrated in earlier studies. Thus, this relation varies from one adhesive to another. Nevertheless, it will be necessary to carry out further tests in order to confirm that point.

   The emission velocity method proposed to CEN is based on the work of Hoetjer (5). This method consists in drawing a straight line through experimental points obtained by plotting on the ordinate the reciprocal value of the formaldehyde concentration, obtained at $22^{\circ}C$, 55% RH (c: formaldehyde concentration in nitrogen), and on the abscissa the ratio n/a, where n is the air exchange rate per hour and a is the board load factor in the chamber in m^2 board per chamber volume m^3. This should yield a correct prediction of the emission for all the n/a values. The curves obtained from boards after 4 weeks of conditioning are shown in Figure 2.

   We have to note that the two melamine-urea-formaldehyde boards do not satisfying this theory. This difficulty excepted, the curve family obtained fits without fault. However, we can say that for a loading rate of 0.5, near that used in the foregoing test, we should obtain a similar ranking, in spite of an inversion between two panels. However, a correlation factor between the two gas analysis methods does not exist, because the values are as follows:

Board F1   Ratio = 1.40
Board R2   Ratio = 1.51
Board R3   Ratio = 1.67
Board R1   Ratio = 2.78

A further experiment with 4 boards was made, Table III. Two series of measurements were carried out on these boards: One was the application of the CEN draft straight line method, and the second were air level measurements in small experimental houses, where the boards were used in roof soffits, as it would be in practice.

   At this point of the study it is not possible to improve our knowledge of the emission trends with this method. However, given that the formaldehyde emission from a particleboard must decrease with time, we decided to measure this effect. Two sets of experiments were carried out parallel to each other for one year at $23^{\circ}C$ and 65% RH, 80% RH, or 30% RH. The boards were tested at regular intervals by both the perforator method and the gas flow

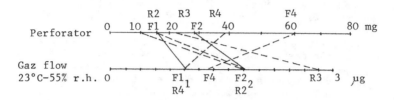

Figure 1. Comparison of board classification according to the
perforator method and the gas analysis method.

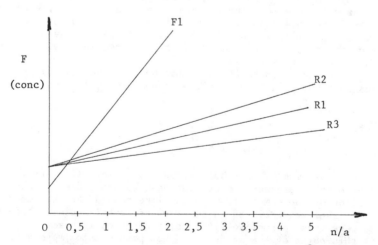

Figure 2. Ratio of the gas analysis methods at 23°C; 55% RH.

Table III. Correlation Between Laboratory and Field Measurements

Board	Content calculated from the CEN method mg/air	Content measured in Experimental homes
R1	0.231	0.079
R2	0.246	0.217
R3	0.239	0.196
R4	0.306	0.209

method at 60oC, 3% RH. Results are shown in Figures 3 and 4. Two distinct phenomena are observed:

1. A decrease with time, together with a certain stabilization after three to four months in a low humidity atmosphere, but results were incoherent, and

2. An abrupt drop in releae under high humidity condition, the three boards proning to give the same results after one year.

Since the influence of ambient air humidity is very significant, we extended this study. Two tests were carried out on boards that were first fully conditioned in a chamber at 23°C and 55% RH: First, short term variations were studied over a period of 1 week, and, second, continuous measurements were taken during sudden moisture uptakes. The results are as follows:

Table IV. Influence of Short-term Moisture Variations

Resin	Perforator (mg HCHO/100g board)				Gas flow (mg/kg board hr)			
	Start	12 weeks 20°C 65%RH	+1 wk 23°C 85%RH	+1 wk 23°C 30%RH	Start	12 weeks 20°C 60%RH	+1 wk 23°C 85RH%	+1 wk 23°C 30%RH
F2 1981 UF	28	21	27	16	8.9	2.3	2.0	2.4
F3 UF	73	53	69	55	3.8	4.7	3.5	8.2
F4 1981 UF-Melamine	64	71	54	57	10.9	4.5	4.7	5.8

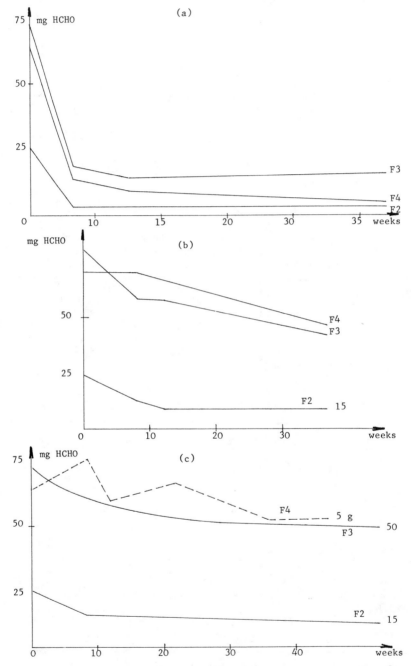

Figure 3. Perforator method. Samples conditioned at: (a) 25°C;
85% RH; (b) 25°C; 30% RH; (c) 20°C; 65% RH.

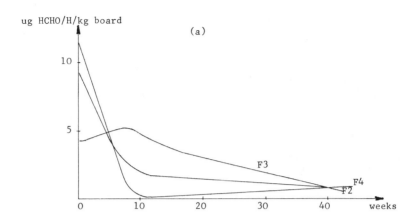

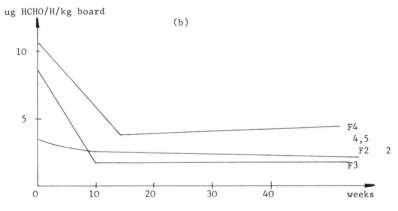

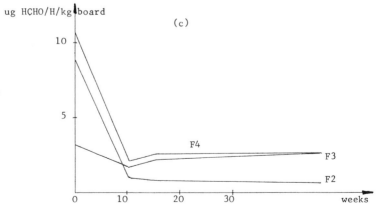

Figure 4. Gas flow method. Samples conditioned at (a) 25°C; 85% RH; (b) 20°C; 65% RH; (c) 25°C; 30% RH

One week variations: After 3 months of stabilization, the boards
are placed one week in a 23°C, 85% RH chamber, then another week in a
23°C, 30% RH chamber. The formaldehyde content was determined at the
end of each period with the perforator and the gas flow at 60°C, 3%
RH. The results are shown in Table IV. The passage into wet
conditions increases the content measured with the perforator for
urea-formaldehyde boards, but decreases it for melamine-urea-
formaldehyde boards. On the other hand, with gas flow at 60°C, there
is an increase only for one urea-formaldehyde board in dry conditons.
      Continuous Measurements: In order to be as close as possible to
practical conditions, measurements were conducted at 23°C. The
nitrogen moisture content was suddenly increased during the test and
the variations of the formaldehyde emission were observed at the same
time.

      We note at each increase of the moisture of the gas stream an
increase of the formaldehyde release, more or less marked according
to the boards. If the same moisture content is maintained, there is
a stabilization of the release, sometimes a decrease of the level,
after 3 or 4 hours. No more variations are observed afterwards
during 36 hours, the maximum duration of our test.
      The mean values of the results obtained during this stabilization
period are given in Table V. We can see a strong increase for one
urea-formaldehyde glue and a lesser increase or nothing at all for
the other glues.

Table V.   Mean Formaldehyde Emission with Variable Humidity
           (HCHO/hour; Nitrogen at 23°C, 20 L/hour)

Board	Nitrogen humidity 55%	80%	Relative increase of the emission from 55% to 80% RH
F1	0.98	0.86	0
F2 1983	2.03	2.93	+45
F4	2.28	2.18	0
R1	1.08	2.39	+121
R2	1.39	3.91	+181
R3	1.87	2.72	+45
R4	2.04	3.79	+86

Summary

The conclusion is simple: It is not possible to predict at the
present state of knowledge, without errors, the risk in practice of
formaldehyde emission from any particleboard for any use by means of
only one simple laboratory measurement. As a matter of fact we find
that:

1. In the case of the analysis of "total" formaldehyde by the perforator method, the ratio of perforator content/emission differs notably from one board type to another, even though for a given type of board, known wood species and glue, the./comparison is valid. The latter condition holds only in the case of quality control during manufacture, and not in product use.

2. In the case of emission measurement, more realistic, it is necessary to take into account the board emission as a function of time. Measurements should be done only after stabilization, of several months if necessary.

3. The great sensitivity to pressure variations may produce sudden and immediate increase of the emission. For example, a consumer taking a shower may experience a blast of formaldehyde release from the shower stall. Thus it is necessary to take into account the final use of the board. A board which releases little formaldehyde at 65% RH may double emission instantaneously when the relative humidity increases to 85%. This type of board should not be recommended for uses in which there is a risk of moisture intake, such as bathrooms or kitchens.

This is to say that in practice for a given particleboard, we have to determine the emission at 23oC and 55% RH, and the influence of humidity in order to classify the board for the appropriate use category.

Acknowledgment

This paper was edited for the ACS Proceedings format by M. M. Kapsa.

Literature Cited

1. Mayer, J. In "Spanplatten - Heute und Morgen"; Weinbrenner, Ed.; DRW Verlag: Stuttgart, 1978, p. 102.
2. "Particleboard-Determination of Formaldehyde Content-Extraction Method Called Perforator Method," European Standard EN-120-1982, European Committee for Standardization, Brussels, 1982.
3. "Particleboards - Determination of Formaldehyde Under Specified Conditions. Method Called: Formaldehyde Emission Method," European Committee for Standardization, CEN Situation Report N 76 E, Brussels, 1984.
4. Determination of Formaldehyde by Gas Analysis, DIN Standard 52 368, 1984. Beuth Publishers, Berlin, 1984.
5. Hoetjer, J. J; Koerts, F. Holz Roh-Werkstoff, 1981, 39, 391.
6. Korf, C. Institute for Surface Technology, Haarlem, Holland, unpublished data, 1985
7. Roffael, E. "Formaldehydabgabe von Spanplatten und anderen Werkstoffen"; DRW Publishers: Stuttgart, 1982.

RECEIVED September 20, 1985

15

Tannin-Induced Formaldehyde Release Depression in Urea–Formaldehyde Particleboard

F. A. Cameron and A. Pizzi

National Timber Research Institute, Council for Scientific and Industrial Research, P. O. Box 395, Pretoria, South Africa

Addition of tannin extract to UF resins in particleboard appear to decrease HCHO-emission over periods of time proportional to the amount of tannin added. The addition of tannin extract appears only to be a "stop-gap" short- to medium-term measure because, once the capability of the tannin to absorb and react with HCHO fumes slowly released by the board has been exhausted, the board revert to emissions similar to those of the UF controls.

The emission of formaldehyde fumes from particleboard manufactured using urea–formaldehyde resins, and its decrease, have now been topics of interest in the timber and wood adhesives industry for a long time. Many solutions, some very effective, to this problem have already been advanced by many authors. In this brief article we do not pretend to present yet another successful or less successful method to control HCHO emission but to show the decrease in the amount of formaldehyde emitted by UF-bonded particleboard, over a period of time, to which tannin extract has been added in small amounts. Tannin extract is an inexpensive commodity in Southern Africa as well as in many other countries in the southern emisphere such as Brazil, Argentina and New Zealand. The method presented, if not completely effective may be an inexpensive system of control of HCHO emission over a limited period of time.

Experimental

Duplicate 12 mm thick three layers particleboard 600 mm x 300 mm in dimensions were prepared in the laboratory using 7 % UF resin solids total on oven dry pine wood chips. The glue mix used was as follows:

UF resin 64 % solids	100	parts by mass
Water	50	parts by mass
NH_4Cl	1.6	parts by mass
NH_3 25 % solution	3.8	parts by mass

0097-6156/86/0316-0198$06.00/0

No wax emulsion was added to the board to avoid the introduction of another factor that could have limited formaldehyde emission.   To this glue mix were added 2 %, 5 % and 10 % UF resin solids by mass of commercial mimosa (wattle, Acacia mearnsii formerly mollissima) bark extract, a commercial flavonoid-type tannin extract.

In another series of panels instead 10 %, 20 % and 50 % on UF resin solids, by mass, of the same extract in spray-dried powder form were added, directly to the wood chips in the glue blender during spraying with the UF glue mix.   The boards were pressed at 25 kg/cm², 170 °C for 7½ minutes with a total cycle of 2 minutes + 2½ minutes + 3 minutes.

One month after pressing, the boards were cut and triplicate samples for each duplicate board   tested according to the dessicator method, using Purpald solution and a colorimeter, for formaldehyde emission over a period of 24 hours and 30 minutes Purpald development. After this initial assessment the samples were placed in a laboratory fan-exhaust oven at a temperature of 50 °C to accelerate the test for a period of two months.   The samples were tested at regular intervals of three weeks over the two months period.   The formaldehyde emission results obtained are shown in Table I.

A further experiment was carried out.   Industrial boards in which 1.5 % tannin extract was added in the glue-blender (1.5 % on UF resin solids) were pressed at 160 °C, 5½ minutes, 25 kg/cm², 9 % UF solids surfaces, 5½ % UF solids in core.   Thickness was of 18 mm finished board.   Average density was of 0.670 g/cm³.   A set of UF controls was pressed under the same conditions.   The results obtained for formaldehyde release are shown in Figure 1 expressed as

$$\frac{\Delta \text{ formaldehyde}}{\Delta \text{ time}}$$

in function of time (in hours) using the dinamic flow method.

Discussion

It is evident from the laboratory experiments that addition of tannin extract to the UF glue mix does not improve the long-term emission of HCHO from the board unless as much as 10 % tannin extract is added.   This may be ascribed to the fact that tannin available to $-CH_2OH$ groups of the UF resin in the glue mix rapidly react with them and thus cannot function as a scavenger of HCHO vapour after the board has been pressed.   The 10 % level is also not too certain as the amount required may vary with pressing temperature, pressing moisture, moisture in the environment after pressing, etc.   It is interesting to note that after one month at ambient temperature the boards with tannin extract added to the glue mix all present lower emission than the UF control.   However, this effect should not last long, even at ambient temperature, as shown from the results of the 50 % accelerated test.

More interesting are the cases in which the tannin extract was added to the chips rather than to the glue mix.   The effect here is also a depressed formaldehyde emission.   The effect appears also to last much longer due to the higher amount of tannin added.   (It must be borne in mind that pure tannin-formaldehyde commercial boards

Table I. Formaldehyde Emission Results of Boards "Spiked" with Tannin Extract

Case No.	Board type	HCHO emission (x 10^{-6}) Accelerated test at 50 °C after				Boards ave density (g/cm³)	Boards ave I.B. (MPa)	Boards ave 24h swelling (%)
		1 month after manufacture	3 weeks	6 weeks	9 weeks			
1	UF control	15.1	5.1	1.21	0.71	0.600	0.46	24.9
2	UF + 2 % tannin in glue mix	10.8	5.5	1.82	-	0.570	0.33	26.5
3	UF + 5 % tannin in glue mix	12.6	7.0	1.85	0.84	0.613	0.42	19.5
4	UF + 10 % tannin in glue mix	11.0	5.4	1.16	0.60	0.570	0.45	20.3
5	UF + 10 % tannin powder	9.2	5.6	1.16	0.65	0.611	0.47	19.5
6	UF + 20 % tannin powder	11.5	3.8	0.76	0.57	0.560	0.52	19.1
7	UF + 50 % tannin powder	9.9	1.9	0.71	0.57	0.551	0.43	16.7

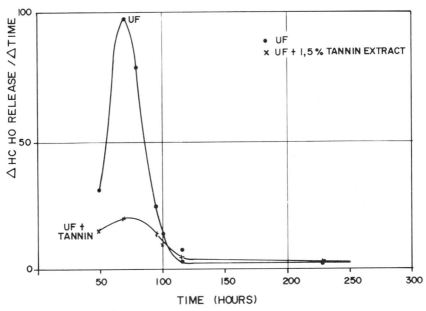

Figure 1.   Differential plot formaldehyde release

manufactured in South Africa, present emissions lower than 0.01 ppm).
The industrial experiment (see Figure 1) shows two results of
interest, namely:

1. that an addition of 1.5 % causes an initial decrease in the
    amount of HCHO emission but that after $\pm$ one week the scavenging
    ability of the small amount of tannin has been exhausted, and
2. that for some reason unknown to the authors, the amount of HCHO
    released in both UF and UF + tannin extract boards abates 70 to
    75 hours after manufacture.

The first point indicate clearly that the addition of tannin is only
a "stop-gap" measure to decrease HCHO-emission from UF-bonded
particleboard as the period of lowered HCHO emission is directly
proportional to the amount of tannin extract (or better of phenolic
matter in the tannin extract = $\pm$ 80 %) added. Once the tannin has,
over a period of time all reacted with the HCHO slowly released, the
board will revert to the same levels of emissions which would have
been obtained without tannin addition. Furthermore, tannin extract
additions of the order of 10 % to 50 % are necessary for longer term
effect. However, notwithstanding the fact that addition of tannin is
only a short to medium term measure it may well constitute a solution
for UF boards which are used for only a limited period of time, such
as in temporary buildings. The other physical properties of the
boards so produced are actually slightly improved by the addition of
tannin (as expected, see Table 1). Small additions of tannins may
also be used, however, to decrease HCHO emission in the factory
during board pressing.

RECEIVED January 14, 1986

16

Effect of Diffusion Barriers on Formaldehyde Emissions from Particleboard

Per Hanetho

Dyno Industrier A.S., Lillestrøm Fabrikker, P. O. Box 160, N-2001, Lillestrøm, Norway

After a discussion of mechanisms for the
liberation and subsequent emission of formaldehyde from particleboard, methods to
assess the extent of these processes are
described. Data are presented for the
formaldehyde emission from particleboard
with various surface treatments. These
data were obtained by a laboratory method
and by large climate chamber measurements
and show that some of the surface treatments studied constitute very efficient
diffusion barriers and considerably reduce
the formaldehyde emission rate.

In this presentation the term "diffusion barrier" will
be used for finishes or overlays for particleboard that
increase the diffusion resistance of the particleboard
surface, thus retarding the rate of mass transfer
(formaldehyde emission) from the board to the surrounding air.

Sources of Formaldehyde in Particleboard

Formaldehyde is liberated during the condensation
reactions that take place when the urea formaldehyde
resin binder in particleboard is cured by hot pressing.
Some of this formaldehyde is retained in the board and
is available for subsequent emission to the surroundings.
    In theory there are several possible states in
which this retained formaldehyde may exist, viz.:
- as monomeric formaldehyde entrapped in voids or adsorbed to the wood
- as monomeric formaldehyde hydrogen-bonded to the wood

- as polymeric (solid) formaldehyde
- as loosely bound formaldehyde, e.g. methylol end
  groups on the resin chain, which readily splits off in
  hydrolytic reactions.

So far no one has been able to demonstrate beyond
doubt in which of the above states the formaldehyde
actually exists. However, at the 4th Annual International Symposium on Adhesion and Adhesives for Structural
Materials in Pullman, WA, September 1984, George Myers
presented a paper concluding that "most of the formaldehyde in a board is chemically, not physically bonded to
resin, to wood, to itself as a polymer, or to ammonia"
(1). He also claimed that all those formaldehyde states
are potentially hydrolyzable, and the more moisture-
sensitive of them, in his opinion, undobtedly act as
sources of a board's emitted formaldehyde. It is, however, not possible to distinguish between formaldehyde
produced from the various states.

Some authors claim that subsequent hydrolysis of
the resin itself also contributes to the formaldehyde
emission. This is not likely, among other things
because the formaldehyde emission is not accompanied by
the bond deterioration and strength loss that would be
the result of resin hydrolysis.

During the manufacture (hot pressing) of the
particleboard the formaldehyde is concentrated in the
core of the board. Tests run on laboratory made
particleboard with the same binder level throughout the
board, have shown about 75% higher content of extractable formaldehyde in the core than in the face (2).
Emission tests indicate an even greater difference
between the two layers of the board.

The concentration gradient that exists between the
core and the face, leads to a migration of formaldehyde
to the surface of the particleboard. From the surface
layer it is released to the surrounding air.

Formaldehyde Emission

The concentration of formaldehyde in the air of a room
containing particleboards, will depend on the content of
formaldehyde in the boards and on the rate of its release. The formaldehyde content of a particleboard is
determined by the binder used to manufacture the board
and a number of production parameters. The release rate
is affected by the temperature and the relative humidity
of the surrounding air, but also by some of the physical
properties of the board. The most important one probably is the diffusion resistance of the surface layer,
which may be expressed by means of a mass transfer
coefficient.

A. Berge et al. (3) and J.J. Hoetjer (4) have
developed models for the formaldehyde emission from
particleboard which can be presented as follows:

$$C_s = \frac{k_g \cdot \alpha}{n + k_g \cdot \alpha} \cdot C^*$$

where
C_s = steady state formaldehyde concentration of the air in a ventilated system, mg/m^3
C^* = equilibrium formaldehyde concentration of the air in an unventilated system, mg/m^3
k_g = mass transfer coefficient, m/h
α = particleboard loading, m^2/m^3, and
n = ventilation rate, h^{-1}.

If the mass transfer coefficient is sufficiently low, the emission will be so slow that the ventilation can manage to remove the formaldehyde at almost the same rate as it is liberated, resulting in a very low form-aldehyde concentration in the air. This presentation deals with what can be achieved in terms of reduced mass transfer coefficient and emission rate by applying some sort of diffusion barrier to the surface of the particleboard. The diffusion barriers studied comprise overlays or surface finishes commonly applied when particleboard is used as a building material, such as wall paper, painting and floor covering, but even over-lays that are used by the furniture and joinery indu-stries, such as veneers, melamine facing and resin saturated paper foils (finish foils).

Test Methods for Formaldehyde Content and Emission

A large number of test methods have been introduced for the determination of the tendency of particleboard to release formaldehyde. Some are analytical methods for the content of formaldehyde in the board, some are emission tests, and some are combinations of the two types. It seems to be generally accepted that the emission tests are the more meaningful ones, among other things because most formaldehyde regulations limit the permissible content of formaldehyde in the air rather than in the particleboard.
It is important to distinguish between those emission tests that measure the emission in a closed, or unventilated, system and those that measure in a venti-lated system. If a particleboard is kept in an unventi-lated system, the formaldehyde concentration will in-crease until it levels off at an equilibrium concentra-tion which will depend on the formaldehyde content of the board under test, the temperature and the relative humidity. The particleboard loading, on the other hand, will not influence the equilibrium concentration, just the time it takes to reach it. The time to reach the equilibrium concentration is also influenced by the mass

transfer coefficient, or in other words by the diffusivity of the surface layer of the board.

In a ventilated system the exhaust air will remove
some of the emitted formaldehyde, and a steady state
concentration will be established. The steady state
concentration will be lower than the equilibrium concentration. How much lower, will depend on the ventilation rate, the particleboard loading and the mass transfer coefficient.

Dyno has contributed to the development of a method,
named the Bell method, for the quantitative determination of the formaldehyde emission from a panel surface
(5). A glass flask or bell having a plane flange is
placed on the surface to be measured. A tight sealing
between the flange of the bell and the panel surface is
very important. The air can be kept in circulation by
means of a membrane pump, pumping about 2 liters per
minute in a closed loop, which also contains a gas
burette. After a predetermined time the formaldehyde
concentration of the air in the gas burette is determined
by a sensitive analytical method.

The Bell method can be used to determine the equilibrium concentration of formaldehyde, C* in the model
above. When the formaldehyde concentration in the Bell
system is plotted against time, the initial slope of
the resulting curve can be used to determine the mass
transfer coefficient, k_g in the same model.

Thus, although there is no air exchange between the
glass bell and the surroundings, the Bell method can be
used to provide data to calculate the steady state concentration in a ventilated system.

Experimental Work

The objective of our work was to determine the effect of
some common surface finishes and overlays on the formaldehyde emission from particleboard. Finishes used in
the building trade as well as such used in the furniture
and joinery industries were studied.

The project plan involved the use of the Bell
method to determine the equilibrium concentration and
mass transfer coefficient for a number of particleboard
samples with different surface finishes and overlays.
The equilibrium concentration and the mass transfer
coefficient were then used to calculate the steady state
concentration in a system with air exchange with the
surroundings, using the model presented above. Tests in
a 24 m^3 climate chamber, in which temperature, relative
humidity and ventilation rate could be varied, were run
to check the agreement between the calculated and
measured values.

Even if it would have been highly desirable to
combine the formaldehyde measurements with determinations of the diffusivity of the various overlays and

finishes, we had to refrain from this. We have, however,
a semi-quantitative conception of the diffusivity of the
coatings and finishes used. We know for instance that
the dispersion paint has a vapour permeability at least
twice as high as the alkyd paint. Also, vinyl surfaced
wall paper has a lower diffusivity than normal wall
paper, and the heavier vinyl materials and paper plastic
laminates are generally considered as being almost im-
permeable.

A surface finish or an overlay may:
1. Affect the equilibrium concentration in an unventilated system, C^*. A coating containing a formaldehyde scavenger would act by binding formaldehyde,
   thus reducing the equilibrium concentration. On the
   other hand some surface finishes will introduce extra
   formaldehyde, and may thus increase C^*.
2. Reduce the mass transfer coefficient, k_g, i.e. the
   rate of formaldehyde transfer from the particleboard
   surface into the room air, without C^* being affected.
   This mechanism is likely for coatings and overlays
   which present a physical restriction to the formaldehyde diffusion, but do not react with formaldehyde.
3. Affect both C^* and k_g. This would be the case for
   finish foils. These are urea or melamine resin
   saturated paper foils which are bonded to the panel
   with urea adhesive. Another example is acid-curing
   lacquers which contain formaldehyde and, at least for
   a limited period of time, substantially increase the
   emission potential, but at the same time is an efficient diffusion barrier for the formaldehyde from the
   particleboard underneath.

It would lead too far here to describe in detail
the various surface treatments studied. Information
about type of material, application methods, adhesive
types, etc., is, however, available.

Results

Table I shows the results obtained with surface finishes
that are common in the building trade.

Discussion of the Results

It should be emphasized that the values presented apply
to the particular materials that we studied, and that the
absolute values cannot be considered as generally valid.
We believe, however, that they can serve to illustrate
the relative reductions in formaldehyde emission that can
be achieved.

Table I

Type of finish	Mass transfer coefficient k_g, m/h	Equilibrium conc., 20°C, C*, mg/m^3	Steady state conc.*), C_s, mg/m^3	
			Calc.	Measured
None (reference)	0.65	2.18	1.69	1.70
Alkyd paint	0.18	0.25	0.10	0.11
Latex paint	0.23	1.98	0.97	1.37
Wall paper	0.24	1.88	0.93	1.67
Vinyl wall paper	0.11	0.39	0.11	0.27
Needle felt carpeting	0.04	0.60	0.065	
Cushion floor	0.04	0.40	0.045	
Carpeting w/foam backing	0.06	0.50	0.088	

*) At 22°C, 60% R.H., ventilation rate 0.5 h^{-1}, particle-board loading 1.6 m^2/m^3

Table II gives the results obtained with overlays that are commonly used by the furniture and joinery industries.

Table II

Type of finish	Mass transfer coefficient k_g, m/h	Equilibrium conc., 20°C, C*, mg/m^3	Calculated steady state conc.*), C_s, mg/m^3
None (reference)	0.40	1.06	0.60
Melamine faced (short cycle)	0.06	1.55	0.25
Paper plastic laminate	0.06	1.19	0.19
Finish foil, 100 g/m^2	0.10	2.01	0.49
Finish foil, 50 g/m^2	0.10	2.80	0.69
Veneer (0.9 mm teak face 1.2 mm pine back)	0.16	0.98	0.33

*) At 22°C, 60% R.H., ventilation rate 0.5 h^{-1}, particle-board loading 1.6 m^2/m^3

The first two measured values are in excellent
agreement with the corresponding calculated values,
whereas for the remaining values the agreement is not
equally good. The most likely reason for this is
inaccuracies in determining the mass transfer coefficients.

The finishes in table 2 were not tested in the climate chamber, because the necessary equipment for the
controlled application of them to full-size particleboards was not available.

Conclusion

Finishing or overlaying particleboard can be an efficient
way to reduce the formaldehyde concentration of the air
in rooms where particleboards are used e.g. as building
panels or in furniture.

Our work shows that all the finishes and overlays
that we have tested, reduce the mass transfer coefficient and lower the rate of formaldehyde emission.

Some of the overlays that are common in the wood-
working industries involve the use of a formaldehyde-
based adhesive. In such cases the adhesive can increase
the emission potential so that, at least for a period of
time, some of the gain due to a reduced mass transfer
coefficient is lost.

Literature Cited

1. Myers, George. Unpublished data.
2. Hanetho, P. Proc. 12th Symp. on Particleboard,
   Washington State University, Pullman, WA, 1978,
   275-286.
3. Berge, A.; Mellegaard, B.; Hanetho, P.; Ormstad, E.B.
   Holz Roh- u. Werkst., 1980, 38, 251-255.
4. Hoetjer, J.J. Holz-Zbl., 1978, 120, 1836-1838.
5. Berge, A.; Mellegaard, B. For. Prod. J., 1979, 29 (1),
   21-25.

RECEIVED January 14, 1986

European Formaldehyde Regulations: A French View

D. Coutrot

Centre Technique du Bois et de l'Ameublement, 10 avenue de Saint-Mande, 75012 Paris, France

Limiting formaldehyde levels should not be set by regulation unless adequate measurement methods are available, except in case of acute health risk. However, it appears that limiting values are being proposed in several European countries, even though we know that it is still difficult to measure and enforce the proposed standard levels and even though the proposed measurement methods have been challenged. In France we want to be certain that we can enforce a standard before we finalize methods and set specific values. Therefore, we still continue to work towards a better understanding and definition of the formaldehyde emission process.

In the present world, one of the key notions of our century is the environment. The environment has become a subject of constant attention for modern man, and it has become a focus of our life and welfare. After having ignored - and even rejected - the environment during the industrial and economic development of the last centuries, we presently incline towards increased respect of nature.

However, it appears that we are changing from one extreme to the other and, instead of striving for harmony between the environment and human welfare, some people reject all that is industrial and demand legislation that is increasingly rigid and prohibitive. The apparent goal is to eradicate any potential aggressor against the environment by legislative means.

Formaldehyde, a strong irritant, is considered one of these aggressors. Since it is a well defined chemical, it has become an easy target for elimination. However, we should remember that formaldehyde is not only an industrial chemical, but is omnipresent in nature: Formaldehyde is present in traces in the living organism where it plays an important part in the metabolic cycles (biosynthesis of the puric nucleus). We can find it in apples, onions, etc. It was also one of the first organic compounds discovered in interstellar space. In fact, in the direction of

0097–6156/86/0316–0209$06.00/0
© 1986 American Chemical Society

Sagittarius, there are two formaldehyde clouds with a mass equivalent
to about one million times the mass of the sun.
On the other hand, formaldehyde is a byproduct of human
activities. It is a combustion product; it is in cigarette smoke, in
wood combustion, and in natural gas flames. Urban air contains
between 10 and 1,000 mg/m^3 of aldehydes, depending on location.
Typical concentrations are shown in Table I:

Table I. Formaldehyde Concentrations in Urban Air

City	Date	Daily Ave. (ppm)
Los Angeles	1961	.005 - .16
	1966	.050 - .12
	1969	.002 - .136
	1979	.002 - .015
New Jersey	1977	.0038 - .0066
Switzerland	1977	.0093 - .01
Federal Republic of Germany	1979	.0001 - .0065
Tokyo	1979	.006 - .17

Formaldehyde is also released from aminoplasts and their derivatives,
such as urea-formaldehyde foam insulation (UFFI), wood adhesives, and
textile finishing agents. It is this supplemental, industrial source
of formaldehyde that has become the subject of risk analysis. Should
we allow products that serve our daily comfort to alter our
environment by releasing an irritating vapor with a pungent odor?
I, for one, believe that comfort alone does not justify such a
situation.
        Another problem with formaldehyde is that we are not yet certain
at which air levels formaldehyde is toxic and dangerous, and at which
levels it causes allergies or other illnesses. The French
Formaldehyde Institute brought a beginning of an answer by making an
evaluation of the toxicity of this product from experiments carried
out in several countries such as the U.S., Sweden, and the Federal
Republic of Germany (1). In France, formaldehyde is classified in
Table C of the Health Code (2) as a dangerous product, except for
preparations containing a maximum of 5 wt%. Moreover, in the
departmental order dated April 25, 1979, the Labour Department
considered formaldehyde an irritant for concentrations included
between 5 and 30 wt% and toxic for concentrations higher than 30 wt%.
This regulation is valid for formol solution.
        Thus, formaldehyde is to be considered an aggressor, and we must:
        1. Reduce the risk of emission that reaches the consumer, and
        2. Evaluate the risk it presents by measuring its concentration
with methods that yield results as close possible to reality.

In an earlier chapter, Romeis has shown that there is
presently no laboratory method that allows meaningful prediction of
formaldehyde emission from particleboards. Why is particleboard so
important? In Europe, this panel represents the biggest use of
aminoplast resins.

The problem with current laboratory methods is that they only
measure formaldehyde at a single time point under equilibrium
conditions. In contrast, real-life use of particleboard involves
climatic shocks. This was well illustrated by a study at the center
for surface technology in Haarlem (3). Figure 1 shows that changes
in air humidity and temperature greatly and promptly influence
formaldehyde emission. Thus, while laboratory tests allow a
qualitative evaluation of the emission risk, they do not permit
quantitative extrapolation to real-life conditions.

Despite this fact, some governments are now enforcing regulations
that are based on test methods that are not suitable for determining
formaldehyde exposure levels and risks. Thus, some countries have
regulated the formaldehyde content of particleboard, relying on the
perforator method, European Standard Method EN 120) (4) which
theoretically measures the total quantity of free formaldehyde in
particleboard. The current regulatory situation for some countries
is shown in Table II.

Table II. Values of Maximum Emission for 100 g of Board (mg)
Statutory or Recommended values (5)

Country	Actual	Target Values
France	50 CTB-S	30
	70 CTB-H	50
Federal Republic	Class E1 0-10	E1 0-5
of Germany	Class E2 10-30	
	Class E3 30-60	
Netherlands	20-25	
Denmark	25	
Finland	30	
Sweden	40	

In France, one proposal has been to keep the 50 g value of the Centre
Technique du Bois, CTB-S for certifying products and to introduce new
classes of formaldehyde content with values of 10 mg/100 g, 25 mg/100
g, etc.

Another proposed regulatory approach takes into account the
formaldehyde concentration in ambient air. There, two cases exist:
The exposure limit values on workplaces, and the exposure limit
values in housing, which are generally one tenth of the workplace
value, see Table III:

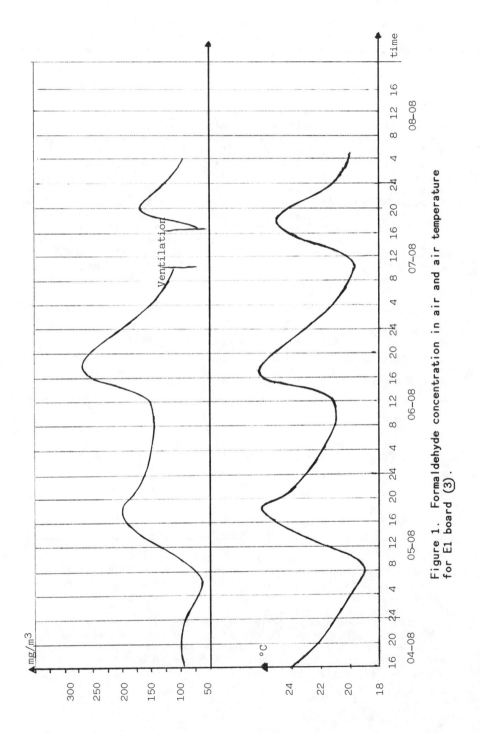

Figure 1.   Formaldehyde concentration in air and air temperature
for E1 board (3).

Table III. Regulation of Formaldehyde Emission in
Various European Countries
(Values in ppm)

Country	Workplaces				Housing	
	1978	1980	1983	Target	1983	Target
France	-	-	2		0.3-0.2 (0.2-0.1)	
Belgium	2	2	2			
Finland	2	2	1	Proposal to put in class 3 (carcinogen)	0.25 0.12[a]	0.1
Denmark						
Old			1		0.12	
New			0.3			
Fed. Republic of Germany	1.2	1.2	1.2		0.1	0.1
Sweden	3	1	1	0.7 old line 0.5 new line	0.4	0.2 (early 1985)
Italy	2	2	2			
Netherlands			2	0.5	0.1	

[a]Houses built after the 1st of January 1983.

The regulation of air concentrations aims at expressing the maximum limit that is admissible. This approach is the most realistic one, because it answers the consumer's legitimate requirements in regard to comfort and health.

The gas flow method would permit the evaluation (under certain conditions) of the risk that we may expect from a board.

Any regulation dealing with the formaldehyde contained in wood products is realistic only if it can be reliably connected to board emission. It seems from our studies (3) that a certain relationship does exist, but this relationship is only valid for boards manufactured on a given factory line. Thus, the relationship between perforator content and gas flow content needs to be more thoroughly studied.

Thus, as we currently try to reduce formaldehyde release into air through regulations, it would seem that actions taken for the sake of "health" are currently going beyond scientifically established facts. Thus, by way of example, in the Federal Republic of Germany the following approach was proposed some time ago: The total formaldehyde air concentration from all sources should not reach air concentrations higher than 0.1 ppm, on and after the 1st of July 1985, and, from the 1st of July 1990, the total concentration in the air should not exceed 0.05 ppm. Fortunately, the latest official

government position does not seem to go towards such an extreme
position. A level of 0.05 ppm is simply not realistic.
    Furthermore, realistic regulations should make possible product
improvement and proper product utilization. In the early 1970s it
was unthinkable to manufacture board with urea-formaldehyde adhesives
having a F/U ratio of 1.5-1.6. Nowadays, it is possible to
manufacture boards of the same quality with glues having a F/U of
1.25-1.2, or even lower.
    Figure 2 shows the evolution of formaldehyde content of
particleboards in Sweden (5,10). Table IV shows relative production
rates of particleboard as a function of formaldehyde emission, using
the sales data for France from CDF-Chimie.

Table IV. French Particleboard Production as a Function of
Formaldehyde Emission (% of total Sales by CDF-Chemie).

Perforator Value	1982	January 1985
About 10 mg/100 g	-	12
About 30 mg/100 g	13	75
About 40 mg/100 g	8	13
Higher than 50 mg/100g	79	-
	100	100

However, we believe that it is of questionable value to demand that
all particleboard sold should be low emitting, because a large part
of the production is sealed and covered before it reaches the
consumer. Thus, French furniture very rarely contains untreated
board, and emission requirements of untreated boards are not a
realistic reflection of emission from the finished product.

Summary

From this short analysis, it emerges that in France we believe in
reasonable reduction of formaldehyde levels, but we do not intend to
engage in rigid formaldehyde regulation, because we believe that:
    1) Current formaldehyde levels are already very much reduced and
do not present a risk at usual current concentrations.
    2) Reducing formaldehyde emission below 0.1-0.2 ppm is currently
unrealistic, because ambient air levels may be higher due to other
formaldehyde sources.
    3) The current methods for measuring formaldehyde emission from
board are expensive, often undependable, and they do not permit a
reliable quantitative extrapolation to real-life conditions at the
present state of research.
    Our view is that one should first establish whether lower board
emission is useful and really necessary under the anticipated board

Value perforator-formaldehyde
mg/100 g

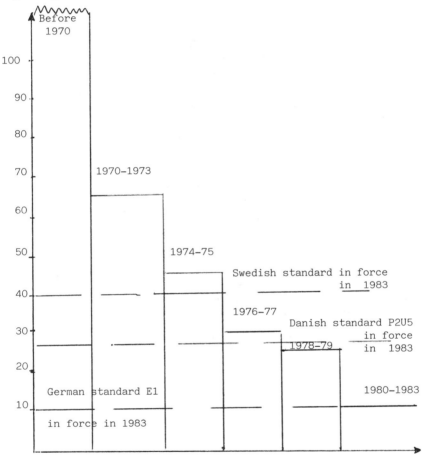

Figure 2. Variation of formaldehyde content of particleboard in Sweden (5).

use conditions before regulations for low emissions are set for all
commercial types of boards.

Acknowledgment

This paper was edited for the ACS Proceedings format by M. M. Kapsa.

Literature Cited

1. "Une Evaluation de la Toxicite du Formaldehyde," Institut
   Francais du Formaldehyd, Paris, 1984.
2. "Valeur Limites d'Exposition aux Substances Toxicque dans les
   Locaux de Travail," Cahier de notes documentaires #106, Institut
   National pour la Recherche et Securite, INRS-ND-1368-106-82,
   Paris, 1982
3. Korf, C. "Etude de quarte types de panneaux de particules
   d'origine francaise," Center for Surface Technology, Haarlem,
   Holland, October 1984.
4. Sundin, B. "Bonded wood panels; Adhesive Systems for the
   Eighties," The World Pulp and Paper Week, Stockholm, April 10-
   13, 1984.
5. "Particleboard-Determination of Formaldehyde Content-Extraction
   Method Called Perforator Method," European Standard EN-120-1982,
   European Committee for Standardization, Brussels, 1982.
6. "Formaldehyde," International Agency for Cancer Research,
   Monograph 29, Geneva, 1984, pp. 347-389
7. Le Botlan, D. "Le Formaldehyde," Laboratoire de Chimie Organique
   Physique, ERA, 315 Centre National pour la Recherche
   Scientifique, 1983
8. Roffael, E. "Formaldehydabgabe von Spanplatten und anderen
   Werkstoffen"; DRW Publishers: Stuttgart, 1982.
9. Johansson, C.E. "Methods for Determining Formaldehyde in Ambient
   Air," FESYP Technical Workshop, Wiesbaden, 1981.
10. Sundin, B. Proc. Int. Particleboard Symposium, 1985, 19, 200.
11. Anonymous; Holz Zentralblatt, February 1, 1985, "Formaldehyde;
    keine Festlegung auf 0.05 ppm."

RECEIVED January 14, 1986

Occupational and Indoor Air Formaldehyde Exposure: Regulations and Guidelines

B. Meyer

Chemistry Department, University of Washington, Seattle, WA 98195

During the past 15 years formaldehyde exposures and emission limits have been significantly lowered. Occuptional threshold limits are now 1.0 ppm or lower in most countries, and actual industrial exposures are almost always half of this value or less. Indoor air standards of 0.1 ppm are now contemplated in several nations, following established procedures for correlating occupational levels of toxic chemicals with ambient air levels. Furthermore, emission standards for UF-bonded wood products have been developed that allow the prediction of formaldehyde levels under various product use conditions before formaldehyde emitting products are installed.

Formaldehyde levels can be regulated by control of air concentrations or by limiting emission at its source. Both approaches are in use. Formaldehyde has been used in pathology labs and hospitals for over a hundred years. It was generally considered a safe chemical, because its pungent odor warned users of over-exposure (1). However, it is well known that some 4% of the population is sensitive to contact dermatitis by formaldehyde (2). This manifests itself in the textile industry and among some consumers who are sensitive to urea-formaldehyde derivatives that are used as finishing agents for ready-to-wear textiles. Problems have been reported especially for shirts, underwear and bed linen.

In the last three decades a special problem arose when large quantities of UF-bonded wood products were used in confined areas that were poorly ventilated. In these applications, several different types of products are often used jointly. Originally, most freshly manufactured UF-bonded products released noticeable quantities of formaldehyde, but emission levels have been reduced by a factor of more than ten (3), and today only defective products, or improperly used products, emit large enough quantities to cause problems. However, the volume of these products has become so large that even a small percenage of complaints can cause a substantial number of complaints. For example, in the U.S. alone, the entire

0097–6156/86/0316–0217$06.00/0

housing stock of seventy million buildings contains at least some of
these products (3).
     Examples of situations that have led to complaints are energy-
efficient homes in Russia, Sweden and Holland; school houses in
Germany, Czechoslovakia and Switzerland; portable temporary offices
and classrooms in Canada and mobile homes in the US. Mobile homes
constitute a special situation, because these residences contain UF-
bonded products in a load ratio of 1.1 m^2/m^3 and recent HUD
regulations allow formaldehyde levels of new homes to reach 0.4 ppm
under standard conditions of 25°C. Such levels are a multiple of
conventional homes. Such levels allow little margin for improper or
defective products, and for emission increases due to warm climates
(3).
     The need for control of formaldehyde emission from UF-bonded wood
products has been recognized since Wittmann (4) reported in 1962 that
extensive use of particleboard in furniture and building envelopes
can cause indoor formaldehyde concentrations exceeding occupational
threshold levels. However, it proved to be difficult to define the
problem because formaldehyde emission from finished products was not
regularly measured, and the correlation between emission rate and the
environmental factors were not yet well established.
     The European particleboard industry (5) led development of
emission testing in the late 1960s. Japan was the first country to
introduce standard product emission testing (6) in 1974. In North
America the failure of the industry to establish voluntary quality
control criteria caused public concern about the safety of
formaldehyde in mobile homes, and problems with poor quality control
of urea-formaldehyde foam emission led the governments of Canada and
the USA to ban the product (7). However, rapid improvement of
products and production quality control have reduced indoor air
levels significantly since the late 1970s when industry and
government jointly commenced work on developing formaldehyde emission
test methods for wood products leading to the HUD standard (8) for
manufactured housing, published 1985, and the development of large
scale air chambers as well as bench-type material test methods.
     Parallel with these devlopments, the energy crisis of 1972 caused
increased emphasis on energy efficient housing. Despite coordinated
action of industry and governments, such as Commercial and
Residential Conservation Programs (9), this led to wide-spread
implementation of poorly understood action, such as reduction of
ventilation to less than 50%, sealing of buildings, reduced heating
that caused moisture condensation problems and accumulation of odor,
including that from unventilated stoves and other human activities.
Thus, large segments of the population rediscovered the importance of
minimizing indoor air pollution, a subject that earlier generations
had learned to optimize hundreds of years ago in order to avoid
tuberculosis (10).

Indoor Air Pollution

Inasmuch as the indoor environment has the purpose to shelter
occupants of buildings, it intrinsically tends to confine indoor
pollutants. Sofar some 300 such pollutants have been identified (10)
and, as mentioned earlier, radon and formaldehyde (4) may reach
occupational threshold levels. Indoor air quality is controlled by a

variety of regulations. The most important are building codes that define building products, ventilation standards, thermal insulation, comfort conditions and similar activities. Other regulations include ambient outdoor standards for criteria pollutants such as sulfur dioxide, nitric oxides and carbon monoxide. Finally, fire codes regulating occupancy rates and smoking regulations also influence indoor air quality.

The depth of current concern for definition and control of the indoor air quality problem is shown by the number of federal agencies that are involved in evaluating and regulating this area in the USA alone. The 16 agencies that form the Interagency Committee for Indoor Air Quality in the US(11) include: the Environmental Protection Agency, (Co-chair), Department of Energy, (Co-chair), Department of Health and Human Services, (Co-chair), Consumer Product Safety Committee, (Co-chair), Bonneville Power Administration, Department of Defense, Federal Trade Commission General Services Administration, Department of Housing and Urban Development, Department of Justice, National Aeronautics and Space Administration, National Bureau of Standards, Occupational Safety and Health Administration, Tennessee Valley Authority, Department of Transportation, and the Small Business Administration.

Determination of Occupational Threshold Levels

The acute toxic effects of formaldehyde are reasonably well known (2). The health effects of formaldehyde have been documented by by NIOSH (12) and OSHA and by a review by the National Research Council for EPA. The setting of standards for formaldehyde has followed the usual standard setting procedure for all toxic chemicals (10). Health effects can be considered to fall into three categories: acute effects, chronic irritation or sensitization, and cancer risk. The well established standards were shaken in 1979 when the Chemical Industry Institute for Toxicology in North Carolina discovered that high formaldehyde concentraions can can cause cancer in rats (13), because such studies have been generally accepted as the basis for determining carcinogenic threshold limits for any type of chemical. Since extrapolation of these findings to human exposure of mobile home residents and textile workers do not clearly exclude potential cancer risk, the corresponding exposure must be reduced, or alternatively, the method for determining cancer risk must be changed for a large number of chemicals (10). Obviously, the impact of the latter approach on regulation of carcinogens would be significant, as would be its impact on industry as well as on consumers.

Occupational Threshold Levels and Exposures

Most countries have established occupational safety limits of about 1 ppm, Table I. In the US the current levels were introduced in 1970 when OSHA was founded. They are based on the 1967 ANSI standard Z-37.16 that was derived from the American Conference of Governmental Industrial Hygienists (ACGIH), set in 1948. However, ACGIH reduced these levels from 5 ppm to 2 ppm in 1983, and in 1976 NIOSH published a recommended 1 ppm level (12). The Chemical Institute of Industrial Toxicology (CIIT) findings that high formaldehyde levels can cause

Table I. Occupational Exposure Limits for Formaldehyde

Country	Type	Value	Nature	Remarks	Reference
Belgium	TLV	2.0 ppm	ceiling		14
Denmark	TLV	1.2 mg/m^3	ceiling	0.4 new	14
Finland	TLV	1.0 ppm	ceiling		14
Holland	TLV	1.0 ppm	8 hr mean		14
	MAC	2.0 ppm			
Italy	TLV	1.0 ppm	ceiling		14
Norway	TLV	1.0 ppm	ceiling		14
Sweden	TLV	0.8 ppm	8 hr mean		14
	MAC	1.0 ppm	ceiling	0.5 ppm	
Switzerland	TLV	1.0 ppm	ceiling		14
United Kingdom	·TLV	2.0 ppm			14
United States					
OSHA	Max.	10. ppm	30 min/day		15
current:	TLV	5.0 ppm	ceiling		
	MAC	3.0 ppm			
proposed:	MAC	1.0 or 1.5 ppm			15
ACGIH	MAC	2.0 ppm	threshold		15
NIOSH		1.2 mg/m^3	30 min ceiling		15
West Germany	TLV	I.0 ppm	ceiling		14

cancer in rats and mice (13), caused a thorough review and revision
of the entire field. This review has not yet come to a conclusion
and the field will undoubtedly remain in flux. The first official
action by eight federal agencies in 1980 was to find that it was
"prudent to regard formaldehyde as posing a carcinogenic risk to
humans" (15). In 1981 NIOSH issued a corresponding intelligence
bulletin (15), and CPSC banned urea-formaldehyde foam insulation (7)
after the Department of Energy was unable to produce an appropriate
material standard (16). However, the ban was overruled in Federal
Administrative Appellate Court (17). As indicated, ACGIH reduced its
level in 1983. The Department of Health included formaldehyde in its
annual report of carcinogens (18), and a consensus workshop was held
to evaluate toxicity (19).
      Subsequently, the Environmental Protection Agency issued an
advance notice of proposed rulemaking indicating its concerns for the
potential risk that formaldehyde might pose to mobile home residents
and textile workers (20), the Office of Manufactured Housing of the
Department of Housing and Urban Development issued standards for UF-
bonded wood products used in mobile homes (8), and in December 1985
OSHA found that "the current permissible exposure limits do not
adequately protect employee health," and it currently seeks public
comments on whether it should reduce its level of 3 ppm to 1.0 or 1.5
ppm (15). Recently observed occupational levels have been summarized
by Sundin (14), Preuss (21), and EPA (20), Table II. It is readily
seen that under normal working conditions occupational formaldehyde
levels are no longer approaching occupational limits.

Table II.   Recently Observed Occupational Exposure Levels

Work Place	Exposure Level (ppm) Mean	Maximum	Reference
US Funeral Homes	0.41	1.7	20,21
Textile Industry	0.25	0.70	20,21
UF Resin Manufacture	0.24	0.59	20,21
Hospital Pathology		0.66	20,21
Plywood Manufacture	0.35	1.2	20,21
Acid cure varnishes	0.94		20
Furniture Manufacture		0.92	21
Fertilizer Manufacture		0.40	21
Foundry Manufacturers		1.2	14,21

Comparison of Occupational and Ambient Air Guidelines

Over the past several decades correlations have been established
between occupational levels and ambient air levels (10). Several of
these rules also hold for indoor air. In a nucleus, the basis for
the correlation is that doses are often additive over time, and that
there needs to be a safety factor for protecting infants and other
sensitive elements of the population. Several countries and agencies
have responded to this uncertainty by setting indoor air formaldehyde
limits. These limits are usually arrived at by modifing the
occupational threshold levels by a factor of ten. This factor is due
to the increase in exposure time, when going from a 40 hr workplace
to a home where one might spend a full 168 hr week, and by adding a
safety factor of about 3 for protecting specially sensitive
indivuals, such as children, old people, and people with pre-existing
sensitivites who could avoid a job involving formaldehyde exposure
but cannot avoid living in their home.
        The additivity of doses derives from time integration, usually
over a period of a week, assuming that dose-response curves are
linear within the corresponding concentration range. Thus, assuming
for example an air level of 1 ppm, industrial workers experience a
weekly dose of:

$$1 \text{ ppm} \times 8 \text{ hr/day} \times 5 \text{ days} = 40 \text{ ppm hrs/week} \qquad (1)$$

In contrast, an infant and a homemaker who, according to worldwide
studies on human activity patterns, spend as much as 20 hrs/day at
home (10), and who live in a mobile home with the same air
concentration as the above worker would experience:

$$1 \text{ ppm} \times 20 \text{ hr/day} \times 7 \text{ day} = 140 \text{ ppm hrs/week} \qquad (2)$$

It is common to express the exposure in weekly time-averaged air
levels. For the above cases the corresponding levels would be
40/168 = 0.24 ppm for the industrial worker, and 140/168 = 0.83 ppm
for the homemaker. In reality, the effect of these exposures will be
modified by many additional factors, such as rest periods (which are
shorter for the mobile home residents than for workers), and
additivity deviations. Thus, general populations are commonly

protected by addition of a safety factor of about 3, which also
includes individual differences in sensitivity.

Indoor Air Levels

Problems arise when unreacted formaldehyde remains in products that
reach less chemically educated and less prepared users in the forest
products industry, and, eventually, consumers who are likely unaware
that they are inadvertantly exposed to residual vapors emanating from
building materials. The most common human response to formaldehyde
vapor is eye blinking, eye irritation, and respiratory discomfort,
along with registration of the pungent odor (2,22,23). The threshold
for registration of formaldehyde strongly differs among people, and
its impact depends on many factors. Thus, some people become
accustomed to what they may consider the natural odor of "wood",
while others become increasingly sensitized. The absolute odor
threshold is 0.05 ppm (24). The dose-response curve for formaldehyde
odor perception among healthy young adults is shown in Figure 1.
Results from recent formaldehyde indoor studies confirm the
observations by Wittmann in 1962 (4) and show that formaldehyde
threshold levels for individual perception are still approached in
many living situations, and are exceeded in certain cases as
highlighted in Table III:

Table III.   Observed Indoor Air Formaldehyde Exposures

Location	Mean Level (ppm)	Reference
Absolute Odor Threshold	0.05	24
Urban Air	0.005	20
Dutch Residenced	0.08	14
Wisconsin Mobile Homes	0.24	25
Minnesota Mobile Homes	0.40	10
Texas Mobile Homes	0.11	26
20 Swedish Homes, 1978	0.3	14
Canadian-UFFI homes	0.065	27
UK-UFFI Building	0.093	22
Conventional Canadian homes	0.034	27
UK Conventional Homes	0.047	22
Bonneville Power Admin.	0.092	28

Several countries and agencies have responded to formaldehyde
complaints by setting indoor air formaldehyde limits. As indicated
above, these limits are usually arrived at by modifying the
occupational threshold levels by a factor of ten. A short summary of
such levels is shown in Table IV:

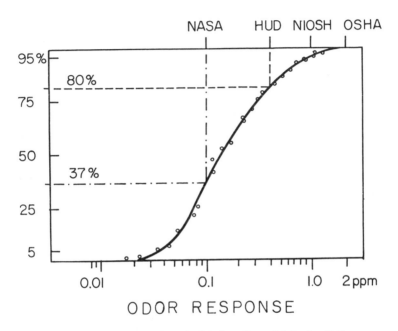

Figure 1.   Odor threshold for formaldehyde (**24**).

Table IV.   Indoor Air Exposure Limit Guidelines or Regulations

Country	Agency or Organization	Level (ppm)	Status	Reference
Denmark		0.12	Law	14
Finland		0.25	Guideline	14
		0.12	1983+	
Holland		0.1	Guideline	14
Italy		0.1	Guideline	14
Sweden		0.4	Guideline	14
USA	ASHRAE	0.1	Guideline	10
	USAF	0.1		10
	USN	0.1		10
	NASA	0.1		10
	Wisconsin	0.6		29
	Minnesota	0.5		30
	HUD, target	0.4	Regulation	8
West Germany		0.1	Guideline	14,31

This area is still in flux. One major problem is that one needs to develop better measurement methods for formaldehyde at low levels, and one needs to have a better field measuring protocol for measuring meaningful formaldehyde levels that are dependent on age of the product, temperature, humidity, and ventilation rate as well as the activities of occupants. All these problems could be reduced, if formaldehyde emission would be effectively controlled at the source. A major effort is now under way to achieve this.

Material Standards for Formaldehyde Emission

The incidence of perceptible formaldehyde in homes, offices and schools has caused widespread uncertainty about the safety of living with formaldehyde. This uncertainty was enhanced by the large scale installation of urea formaldehyde foam insulation (UFFI) because a substantial part of this material was made from small scale resin batches prepared under questionable quality control conditions, and was installed by unskilled operators (10). The only reliable way to avoid such uncertainty is to know the emission rate of products and develop a design standard that allows prediction of indoor air levels. The first and most important step in this direction was achieved with the development and implementation of material emission standards. As indicated above, Japan led the field in 1974 with the introduction of the 24-hr desiccator test (6), FESYP followed with the formulation of the perforator test, the gas analysis method, and later with the introduction of air chambers (5). In the U.S. the FTM-1 (32) production test and the FTM-2 air chamber test (33) have made possible the implementation of a HUD standard for mobile homes (8) that is already implemented in some 90% of the UF wood production (35), regardless of product use.

Table V. Formaldehyde Emission Test Methods

Country	Chamber Test	Production Test	Reference
Belgium		Perforator Value[a]:	34
Class 1		14	
Class 2		28	
Class 3		42	
Danish	0.225 m^3 chamber[b]:	Perforator Value[a]:	14.34
E-15	0.15		
P-25U		average value: 25	
P-25B	0.30	max. 10	
Finland	0.12 m^3 chamber	Perforator[a]:	14,34
		40	14,34
France		50	14
Holland		10 av.; 12 ceiling	14
Japan		24-hr dessicator[c]:	6
Norway		Perforator[a]:	
30	14,31		
Swedish	1 m^3 chamber	40	14,31
Spain		50	14
Switzerland		20	14
United Kingdom		50 average	14,34
United States			
Mobile Homes:	FTM-2 Chamber[e]:	FTM-1,2hr dessicator[f]	8
	1,000-1,200 cft		
Plywood	0.2		8
Particleboard	0.3		8
MDF	0.3[g]		34
West Germany	39 m^3-chamber[h]	Perforator Test[a]:	14,35
E-1	0.12 mg/m^3	10	
E-2	0.12 - 1.2	10 - 30	
E-3	1.2 - 2.75	30 - 60	

[a]: Perforator Test: CEN-Standard EN 120-1982, (34)
[b]: Danish Air Chamber: Load: 2.25 m^{-1}; 23°C; 45 %RH; 0.50 ach
    (currently still 0.25 ach), (14)
[c]: Finnish Chamber: Load: 1 m^{-1}, 20°C, 65 %RH, 0.5 ach, (14)
[c]: Japanese Industrial Standard, JIS-A5908-1977, (6)
[d]: Swedish Air Chamber; CEN Situation Report-1983 (14):
    Load: 1; 23°C; 50 %RH; 0.5 ach, (40)
[e]: HUD air chamber, FTM-2: Load 1.1; 77°F; 50 %RH; 0.5 ach (8)
[f]: NPA-HPMA-FI, FTM-1, 2 hr desiccator test, (32)
[g]: Industry Product Standard, (34)
[h]: ETH standard chamber: Load: 1; 23°C; 45 %RH; 1 ach, (35)

In Europe, the most widely used test method is a CEN standard method
(37), the FESYP perforator test method developed in the middle 1960s
by Verbestel (5). However, this method is no longer sensitive enough
to differentiate among the products in the lowest emission classes,
such as German Class E-1 (35), because it is excessively sensitive to
moisture content of the wood and its findings depend on whether

formaldehyde is determined colorimetrically or by standard iodine
titration. This test is based on the assumption that vaporizable
formaldehyde is fully removed from small samples if they are boiled
in toluene for 4 hours at 110°C. This assumption, while never
theoretically confirmed, and brought into question by work reported
by Romeis in another chapter, has proven a useful basis for
correlation between laboratory tests and actual air levels for
individual products. However, this test is unsuitable for
comparisons of different types of products such as particleboard and
plywood. Another convenient method is the WKI test developed by
Roffael (39), but it also uses elevated temperatures that might
distort product rankings. However, the correlation between these
quality control methods and the air chamber tests has been well
established and is clearly sufficient for complaint investigations.
A summary of currently used methods is provided in Table V.
    The test results can be used to predict indoor air levels if load
factors, ventilation rates, temperature, air humidity and occupant
activities are known. This subject is explained in Chapter 1. By
way of example, Figure 2 shows the safe product range that has been
established in Sweden for particleboard use in conventional housing
(14). As soon as product performance is widely disclosed and
builders and architects become familiar with the product ratings,
formaldehyde complaints will rapidly decrease and likely become a
thing of the past.

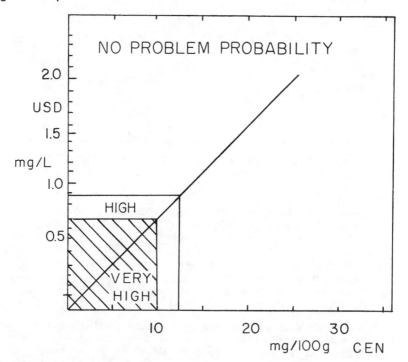

Figure 2. Safe emission limits for UF-bonded pressed wood
products; P = perforator value (mg/100g); USD = desiccator value
(mg/L), after reference 14.

Summary

During the past ten years the occupational and ambient indoor formaldehyde guidelines and regulations have been thoroughly reviewed and revised. The recent development of product emission standards will greatly reduce confusion about the safety of UF-bonded products and will make it possible to eliminate products with unacceptably high emission before they are installed.

Literaure Cited

1. Meyer, B. "Urea-Formaldehyde Resins"; Addison-Wesley Publishers: Reading, MA, 1979.
2. Ulsamer, A.G., Beall, J.R.; Kang, H.K.; Frazier, J.A. Hazard Assessment of Chemicals 1984, 3, 337.
3. Meyer, B.; Hermanns, K. J. Air Pollution Control Assoc. 1985, 35, 816-821.
4. Wittmann, O. Holz Roh- Werkstoff 1962, 20, 221-224.
5. "Analysis Method; Formaldehyde Determination in Air, Photometric Method, and Iodometric Method," Federation of European Particleboard Manufacturers, Giessen, Germany, 1975.
6. "Materials and Fittings, A-5906-1983 Medium Density Fiberboard; A-5907-1983 Hard Fiberboards, A-5908-1983 Particleboard, A-5909-1983 Dressed Particleboard, A-5910 Dressed Hard Fiberboard," Japanese Industrial Standards, (Official English Translation, available through the American National Standard Institute, New York), 1985.
7. "Ban of Urea-Formaldehyde Foam Insulation," U.S. Consumer Product Safety Commission, Federal Register, 1982, 47, 14366-14421.
8. "Manufactured Home Construction and Safety Standard," U.S. Code of Federal Regulations, 1985, 24, Part 3280.406, (U.S. Department of Housing and Urban Development), and Federal Register, Vol. 48, pg 37136-37195, 1983.
9. "Residential Conservation Program," National Energy Conservation Policy Act, Part I, Title II, Public Law 95-619 of November 9, 1978, U.S. Congress.
10. Meyer, B. "Indoor Air Quality"; Addison-Welsey Publishers: Reading, MA, 1984.
11. "Interagency Committee on Indoor Air Quality - Comprehensive Indoor Air Quality Research Strategy," U.S. Environmental Protection Agency, U.S. Department of Energy, U.S. Department of Health and Human Services, U.S. Consumer Product Safety Commission, 1985.
12. "Occupational Exposure to Formaldehyde, Criteria for a Recommended Standard," National Institute for Occupational Safety and Health, 1976.
13. Swenberg, J.A.; Kerns, R.E.; Mitchell, R.E.; Gralla, E.J.; Pavlov, K.L. Cancer Research, 1980, 40, 3908-3402.
14. Sundin, B. Proc. Int. Particleboard Symposium, 1985, 19, 200.
15. "Occupational Exposure to Formaldehyde," Occupational Safety and Health Administration, Federal Register 1985, 50, 50412-50499.
16. "Urea-Formaldehyde Foam Insulation, Interim Standard," U.S. Department of Energy, Federal Register, 1980, 45, 63786, and 1981, 46, 8996.

17. "Gulf South Insulation vs. CPSC," Federal Reporter, 1983, 701,
     2nd ed., 5th circular, 1137.
18. "Third Annual Report on Carcinogens," U.S. Department of Health
     and Human Services, National Toxicology Program, 1981.
19. "Report of the Consensus Workshop on Formaldehyde,"
     Environmental Health Perspectives, 1984, 58, 323-381.
20. "Formaldehyde: Determination of Significant Risk," U.S.
     Environmental Protection Agency, Federal Register 1984, 49,
     21870.
21. Preuss, P.W.; Dailey, R.L.; Lehman, E.S. Adv. Chem. 1985, 210,
     247.
22. Andersen, I.; Mølhave, L., Chapter 14 in "Formaldehyde
     Toxicity"; Gibson, J.E., Ed.; McGraw-Hill: New York, 1983.
23. Mølhave, L.; Bisgaard, P.; Dueholm, S. Atmospheric Environment,
     1983, 17, 2105-2108.
24. Berglund, B.; Berglund I.; Johansson,I.; Lindvall, T. Proc.
     Third Int. Symp. Indoor Air Quality and Climate, Vol 3., Swedish
     Council for Building Research, Stockholm, 1984, pp. 86-96.
25. Hanrahan, L. P.; Dally, K. A.; Anderson, H. A.; Kanarek, M. S.;
     Rankin, J. Am. J. Public Health 1984, 74, 1026-1027, and J. Air
     Pollution Control Assocation 1985, 35(11), 1164.
26. Stock, T. H.; Monsen, R. M.; Sterling, D. A.; Norsted, S. W.
     78th Annual Meeting Air Pollution Control Assoc., Air Pollution
     Control Assoc.: Detroit, 1985.
27. Shirtliffe, C.J.; Rousseau, M.Z.; Young, J.C.; Sliwinski, J.F.;
     Sim, P.G. Adv. Chem. 1985, 161-192.
28. "Preliminary Formaldehyde Testing Results for the Residential
     Standards Demonstration Program," Bonneville Power
     Adminsitration, U.S. Department of Energy, Reiland, P.;
     McKinstry, M.; Thor, P., 1985.
29. "Wisconsin Statutes," 1983, Section X, "Proposed Standard for
     Mobile Homes," State of Wisconsin.
30. "Minnesota Statutes," 1985, Section 144.495, "Formaldehyde
     Product Standard," State of Minnesota.
31. "Formaldehyde, A Joint Report of the Federal Health Agency,
     Occupational Health Agency, and the Environmental Agency," 1984,
     October 9., Federal Agency for Youth, Family and Health, Bonn,
     Germany.
32. "Small Scale Test Method for Determining Formaldehyde Emission
     from Wood Products, Two Hour Dessicator Test, FTM-1," National
     Paricleboard Association, Hardwood Plywood Manufacturers
     Association, Formaldehyde Institute and U.S. Department of
     Housing and Urban Development, Federal Register, 1982, 48,
     37169.
33. "Large Scale Test Method for Determining Formaldehyde Emission
     from Wood Products; Air Chamber Method, FTM-2" National
     Particleboard Association, Hardwood Plywood Association, U.S.
     Department of Housing and Urban Development, Federal Register,
     1982, 48, 37169.
34. "LOFT paneling and Mobile Home Decking," and "Fiberwood .3"
     Weyerhaeuser Corporation, Tacoma, WA, 1981 and 1984.
35. "ETB-Baurichtlinie fuer die Vermeidung von unzumutbaren
     Formaldehydekonzentrationen, Berlin, 1979.
36. Birner, B Wood and Wood Products, 1985, 90(5), 92.
37. "Particleboard-Determination of Formaldehyde Content-Extraction
     Method Called Perforator Method," European Standard EN-120-1982,
     European Committee for Standardization, Brussels, 1982.

38.   "Guideline on the Use of Particleboard with Respect to Avoiding
      Intolerable Formaldehyde Concentrations in Room Air," Committee
      for Uniform Technical Construction, Institute for Construction
      Technology, (ETB), Berlin, translated by U.S. HUD, 1980.
39.   Roffael,E. "Formaldehydabgabe von Spanplatten und anderen
      Werkstoffen," DRW Publishers: Stuttgart, 1982.
40.   "Particleboard-Determination of Formaldehyde Emission under
      Specified Conditions; Method Called: Formaldehyde Emission
      Method," European Standard Situation Report EN-N76E-1983,
      European Committee for Standardization, Brussels, 1983.

RECEIVED January 14, 1986

Author Index

Subject Index

Production by Joan C. Cook
Indexing by Susan Robinson
Jacket design by Pamela Lewis

Elements typeset by Hot Type Ltd., Washington, DC
Printed and bound by Maple Press Co., York, PA